NOT IN OUR BACKYARDS!

NOT
IN OUR BACKYARDS!

Community Action for Health and the Environment

Nicholas Freudenberg

Foreword by Lois Marie Gibbs

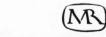

Monthly Review Press
New York

Copyright © 1984 by Nicholas Freudenberg
All rights reserved

Library of Congress Cataloging in Publication Data
Freudenberg, Nicholas.
  Not in our backyards!

  Bibliography: p.
  Includes index.
  1. Environmental health—United States—Citizen
participation—Case studies.   2. Environmental protection—United States—Citizen participation—Case
studies.   3. Environmental health—Citizen participation—Case studies.   4. Environmental protection—
Citizen participation—Case studies.   I. Title.
RA566.3.F74 1984   363.7'0525   84-10524
ISBN 0-85345-653-4
ISBN 0-85345-654-2 (pbk.)

Monthly Review Press
155 West 23rd Street
New York, N.Y. 10011

Manufactured in the United States of America

10 9 8 7 6 5 4 3 2 1

In memory of my father, Albrecht W. Freudenberg, who died unnecessarily as a result of exposure to benzene, and for my son, Sasha Freudenberg-Chavkin, who deserves a healthier and more just world

CONTENTS

Foreword by Lois Marie Gibbs   9

Preface   11

Acknowledgments   13

1. The Corporate Assault on Health:
   Environmental Health Hazards and Their Causes
   Case Histories: Hardeman County, Tennessee;
   Pine Ridge, South Dakota; Elizabeth, New Jersey   15

2. Science and Politics:
   The Limitations of Environmental Health Research
   Case History: Love Canal, New York   42

3. Environmental Regulations: Who Is Protected?
   Case History: Washington Heights, New York   60

4. Arming Yourself for Battle:
   How to Get Information on Environmental
   Health Hazards
   Case History: Staten Island, New York   82

5. Building an Organization that Can Survive
   Case History: Rutherford, New Jersey   114

6. Strategies for Action: Community Education
   Case History: Philadelphia, Pennsylvania   133

7. Strategies for Action: Legal and Legislative
   Case Histories: Southern Utah; Elizabeth, New Jersey;
   Mendocino County, California; Lincoln County,
   Oregon   159

8. Strategies for Action: Community Organizing
   Case Histories Fort Worth, Texas; Warren County, North
   Carolina   180

9. Coalition Building: Issues and Problems—
   Labor/People of Color/Women/Peace Groups/Third
   World Groups
   *Case Histories: Shell Strike, California; Harlem, New
   York; Suffolk County, New York; Pentagon Action;
   Stop Project ELF, Wisconsin*   194

10. Toward a National Environmental Movement
    Case Histories: Japan; West Germany   239

References   273

Resources for Action   291

Index   296

FOREWORD
by Lois Marie Gibbs

There's a new people's movement growing in this country. It's made up of tens, if not hundreds, of thousands of people—blue collar workers, housewives, people of color, farmers, small business owners, homeowners, renters, urban people, and rural people. They're young and old and every age in between. They have mainly low and moderate incomes, though the well-to-do are not immune.

The thing they have in common is a fear and concern about how this country manages hazardous waste. This fear and concern has literally come home for them as they've discovered that hazardous waste problems don't just happen in places with odd sounding names like Love Canal or Times Beach or Stringfellow but can happen in their own backyards. They've found out that they're living near one of the over 100,000 hazardous waste sites that exist across the country. Or they've found that their community has been chosen to receive a share of the 250 million tons of hazardous waste the United States produces every year.

They've become concerned because of the growing awareness that these toxics *hurt* people now, can damage their children through birth defects and harm future generations through long-term genetic change. Hazardous wastes can turn the home that most people have worked so hard to buy into a toxic prison, which they can't afford to sell (even if they could find a buyer) and can't afford to stay in because of what the chemicals are doing to their health.

They've become concerned because none of the people in charge seem to care. The public officials and policymakers seem more interested in keeping people calm and in minimizing the nature of the problem than in doing something about it. The corporations and the hazardous waste handlers only seem to care about profits. Both the industry and the government policymakers

9

seem to react by fighting "dirty," even though these concerned citizens almost always start out being very polite.

So the time comes when concerned community people decide they have to fight. Parents fight to protect their children. Working people fight to protect all of the things they have worked so hard to get. Homeowners fight because they feel they're being backed into a corner. Farmers and business owners fight to protect their lives and their health. And over it all is a fight for justice.

When these concerned community people get angry and ready to fight, they transform themselves into leaders and begin talking to their neighbors. In hundreds of communities, they have created strong people's organizations and have begun to build political power.

And they've used their political power to bring their elected leaders to account and irresponsible corporations to heel. They've borrowed strategies and tactics from all of the social justice movements that have come before and have made up some new approaches of their own. They've knocked on doors, leafletted, held rallies, teach-ins, and protest marches. They've used their economic muscle to launch boycotts and they've used their courage and anger to expose lies and evasive answers from powerful officials and big corporations.

This is a book about people struggling against injustice, fighting for their health, homes, jobs, and children. It is a book about victories that have been won and victories that will be won.

I know these struggles very well. In 1978 I was a Niagara Falls housewife living in a neighborhood called Love Canal. I found chemicals in my own backyard and these chemicals were hurting my children. I tried it the nice way and experienced all of the things the people felt in the stories this book tells. Then I got angry. And then I fought by organizing my neighbors into the United States' first citizens' organization fighting hazardous waste. We won. So have many of the people in this book. So can you. And so can we all.

PREFACE

Every week a new environmental disaster makes front-page news. The U.S. Environmental Protection Agency announces that our grain and citrus products are contaminated with ethyl dibromide (EDB), a cancer-causing fumigant. Health officials in New Jersey find dozens of sites where improper disposal of industrial waste has led to human exposure to dioxin, one of the deadliest chemicals known. Residents of Dallas, Texas, force the government to close a smelter that has been polluting their air with high levels of lead, which causes mental retardation in children.

Why are we facing this continual assault on our health? What can we do to stop it? This book seeks to answer these questions by giving its readers—concerned citizens, environmentalists, students, activists, and others—the information and skills they need to protect themselves, their families, and their communities from environmental health hazards. It begins with a description of how corporations and government agencies contaminate our air, water, food, and soil with toxic substances (chapter 1). It reviews the accomplishments and limitations of two methods used to control human exposure to hazardous materials: scientific research and government regulation (chapters 2 and 3). The next five chapters describe how community residents can take action against threats to their health. They discuss how to investigate a problem (chapter 4), how to organize a group (chapter 5), how to educate the public (chapter 6), and how to pursue legal, legislative, and direct action remedies (chapters 7 and 8). The final chapters propose a plan for strengthening the environmental movement by developing coalitions with other constituencies (chapter 9) and a political program that can unite grass-roots environmental groups and the national conservation organizations (chapter 10). Each chapter begins with one or more case histories that document a particular community's battle to better its health and environment. Analyz-

11

ing the successes and failures of these efforts helps us learn from a decade of environmental activism.

The research on which this book is based comes from several sources. In 1981 I conducted a survey of community organizations involved in environmental health issues. I mailed a questionnaire to 242 groups in 38 states, asking about their goals, strategies, and problems. Forty-five percent (110 groups) answered. My analysis of how to successfully organize a group is based largely on the results of this survey. All quotations in the text that are not foot-noted are taken verbatim from the questionnaires. In addition, I interviewed more than a dozen environmental activists (see the acknowledgments) to obtain more detailed accounts of a number of community struggles. Quotations from these interviews are cited in the reference notes to each chapter.

Not in Our Backyards! is part of an ongoing effort to help people learn from the experiences of the environmental movement. I hope that readers will contribute to this process by sending me descriptions of how they used the book, reactions to—or criti-cisms of—its contents, and accounts of environmental campaigns that they have been involved in (materials should be sent to Box 609, Hunter College School of Health Sciences, 425 East 25th Street, New York, N.Y. 10010). By profiting from each other's successes and failures, we can begin to convert the energy of grass-roots organizing into the political power needed to safe-guard our health and our environment.

ACKNOWLEDGMENTS

Many people helped to make this book possible. More than one hundred activists from around the country took the time to respond to my survey on environmental organizing. Their willingness to describe and analyze their successes and failures provided me with the raw material for the book. I also owe a great debt to those who consented to be interviewed: Edwina Cosgriff of Bring Legal Action to Stop the Tanks (BLAST) on Staten Island, New York; Carol Froelich and Louisa Nicols of We Who Care in Rutherford, New Jersey; Professor David Wilson of Vanderbilt University in Nashville, Tennessee; Nell Grantham, J. R. Taylor, Margaret Williams, and Carol Pearcy of Tennesseans Against Chemical Hazards (TEACH); Helene Brathwaite of the Parents' Association of P.S. 208 in Harlem, New York; Seth Borgos, research director for the Association of Community Organizations for Reform Now (ACORN) in Washington, D.C.; Jacinta Fernandes and Gloria Davis of the Coalition for a United Elizabeth in Elizabeth, New Jersey; George McDevitt of the International Chemical Workers' Union in Nanuet, New York; Heather Baird Barney, research scientist for Massachusetts Fair Share in Boston; Gregor McGregor, environmental attorney in Boston; Susan Sladack, assistant for congressional affairs in the U.S. Environmental Protection Agency's Boston office; and Tom McShane, assistant secretary of environmental affairs, Massachusetts Department of Environmental Quality Engineering in Boston. (The Boston interviews were conducted by my research assistant Carl Blumenthal.)

Many of my students at the Hunter College School of Health Sciences and the Columbia University School of Public Health helped to conduct and analyze the survey and to track down references in the library. I am grateful to Ruthann Evanoff, Lea Cassarino, Millicent Reynolds, Naomi Escalante, De Lys Saint Hill, Ngozi Moses, and especially Ellen Zaltzberg and Carl Blumenthal. My colleagues at Hunter, particularly Gib Shimmel, provided me

13

with ongoing support for this project. The Faculty Research Award Program of the Board of Higher Education and the Professional Staff Congress provided partial funding for the survey.

My friends and comrades at the Washington Heights Health Action Project—Sally Kohn, José Alfaro, and Rosa Garcia—helped me to understand the day-to-day needs of community organizers; it was our collective experience that persuaded me to write this book.

Naomi Nemtzow, Kathryn Shield, and Florence Kalian faced the unenviable task of converting pages of drafts into a presentable manuscript, while Laura Parker successfully persuaded the often recalcitrant CUNY computer to analyze the survey results. I thank them for their work.

Mary Heathcote edited the first several chapters of the book; her comments were always incisive and helpful. I am especially grateful to Susan Lowes, my editor at Monthly Review Press. Her suggestions for reorganization helped to make the book more readable and her support and encouragement got me through the many drafts.

Finally, I want to thank Wendy Chavkin and Susan Schechter, who read through innumerable drafts, always providing useful criticism and suggestions. Their willingness to engage in political discussion helped me to clarify my ideas.

1. THE CORPORATE ASSAULT ON HEALTH:
Environmental Health Hazards and Their Causes

Case History 1: Hardeman County, Tennessee

In 1964 the Velsicol Chemical Company of Chicago bought a 242-acre farm in rural Hardeman County, Tennessee.[1] The company soon dumped 300,000 fifty-five-gallon barrels of chemical waste on the site. The bulldozer that buried the barrels occasionally burst a drum open; its contents disappeared into the soil.

In 1967 a U.S. Geological Survey report showed that the chemicals from the dump site were reaching the wells of the families that lived on a nearby road, but officials took no action. Five years later, the Tennessee Department of Public Health ordered Velsicol to close the dump.

By 1977 residents near the dump were noticing that their drinking water had a foul odor and taste. Among those who lived in the area was Nell Grantham, a licensed practical nurse, and her family. Her neighbors included her parents, her brothers and sisters, and their families. They got together and asked the state health department to analyze their water, but their request went unheeded for six months. When the local health department finally did take samples, they tested only for bacteria. Finding none, the health officer told them their water was safe.

Dissatisfied, Grantham and her relatives brought a sample to the State Water Quality Division laboratories. After making its own tests, the state lab gave the Granthams and their neighbors some grim news: their water contained twelve chemicals including five known carcinogens—benzene, chlordane, heptachlor, endrin, and dioxin. The lab conducted more tests, the results of which were then confirmed by the U.S. Environmental Protection Agency (EPA). EPA officials gave the residents a series of recommendations, including not to drink their water and not to cook with it. They were then told not to bathe in it. Later still they were told that fruits and vegetables should not be grown on their prop-

15

erty, or animals raised there: one resident's hog was killed and sent to the state lab, which found it to be highly contaminated with chemicals. After several months of having water shipped in by National Guard tankers, the residents were connected to the public water system of a nearby town. But before long residents challenged the quality of that water as well, for it too was found to contain chemicals.

Media coverage of the problem forced Velsicol to admit publicly that its abandoned dump might be the cause of the problem. The company agreed to pay for the new hot water heaters, dishwashers, and washing machines that residents had to buy to replace their contaminated appliances. Some residents complained that Velsicol did not live up to this agreement.

Meanwhile, the residents experienced a host of health problems. Skin rashes became common. A child was born with a serious birth defect. Following that birth, none of the couples in the area of the dump were able to conceive a child for more than four years. An environmental health survey by scientists from the University of Cincinnati Medical Center found evidence of liver damage in some of those exposed to the water, and attributed it to the contamination.[2]

While residents were frightened by the ill effects they were suffering, they were even more alarmed about what might happen in the future. As Nell Grantham put it to a journalist, "Who knows? In twenty years my kids may have cancer. I may never have grandchildren. I may not even live to see my kids grown. That's something we don't know. That's something no one knows. But when you've been drinking contaminated water with chemicals that you know can cause cancer, that makes it look worse. My kids are my biggest concern. What kind of life are they gonna have?"[3]

A group of eighty people living close to the dump have filed a $2.5 billion class action suit against Velsicol. Before that, a few residents settled out of court, receiving from $12,000 to $15,000 each from the chemical company. The residents have also demonstrated, petitioned, and written countless letters to local, state, and federal officials. By late 1983, no decision had been made on the class action suit and since both parties vowed to appeal an adverse ruling, it will be years before a final decision is reached. As a result of these efforts, Grantham says, "I may not get but ten cents [from the suit], but they'll know I've been there—the state of Tennessee and Velsicol, too."[4]

On another level, Grantham became a leader of a new statewide coalition, Tennesseans Against Chemical Hazards (TEACH). Composed of eleven community groups across the state, TEACH does public education and legislative lobbying, and supports community organizing against polluters. It is also working for a state tax on chemical companies to be used to clean up abandoned dump sites.

Case History 2: Women of All Red Nations, Pine Ridge Reservation, South Dakota

On the Pine Ridge Reservation in South Dakota, local members of Women of All Red Nations (WARN), an affiliate of the American Indian Movement, became concerned about high rates of miscarriage and birth defects on their reservation.[5] They conducted a household survey and interviewed hospital workers and found that in one month in 1979, 38 percent of the pregnant women on the reservation suffered miscarriages, more than double the national average. A doctor at a local Indian hospital issued a report showing higher than normal rates of cleft palate and other birth defects, and of hepatitis, jaundice, and serious diarrhea among the babies delivered there. Health officials confirmed WARN's suspicion that the reservation had high rates of bone and reproductive cancers.

Over the next two years WARN began an investigation that uncovered a plethora of environmental hazards. It sent samples of water to a private laboratory and found that inadequate sewage treatment had led to fecal contamination of drinking and bathing water. Widespread spraying of pesticides to control grasshoppers and prairie dogs, and of herbicides such as 2,4-D (a component of Agent Orange), had further contaminated the water supply. At WARN's insistence, the Indian Health Service tested for radioactivity in the water and found that several samples exceeded EPA standards. The reservation is downwind from a number of old mines that are surrounded by uranium tailings, a by-product of mining. Furthermore, in the last few years energy companies looking for uranium had drilled 12,000 exploratory holes in the area and many of these drill sites had been improperly capped. And a major spill of radioactive waste in 1962 that had contaminated the Cheyenne River, which flows through the reservation, had never been cleaned up. WARN also suspected that an old air force gunnery range on the reservation may have been used as a dump site for nuclear waste.

Unlike other organizations fighting environmental hazards, WARN is a multi-issue group that existed before the Pine Ridge health hazards were discovered. It was founded in 1978 as a national grass-roots organization of Native American women who wanted to "protect the health, culture, and lives of their families and communities." The group has worked against the use of Indian lands for energy development, against political repression of Native American activists, against sterilization abuse of Native American women, and for improved child care and schooling.

These experiences guided the group's response to the health problems and the environmental hazards. For example, WARN activists called a press conference to announce the preliminary results of the health survey. They demanded an independent study by reputable researchers, emergency shipments of clean drinking water until the existing water supply could be made safe, a system for monitoring water quality, and an end to further energy development in the area.

To support these demands, WARN obtained affidavits from sympathetic health professionals and hired independent consultants to test water quality. Its lawyer used the Freedom of Information Act (FOIA) to obtain more evidence from EPA and the Bureau of Indian Affairs. Appropriate state and federal government officials were besieged with telegrams. On the reservation itself, the group distributed fact sheets describing the problem and listing its demands. The tribal community councils were urged to pass resolutions of support and WARN members met regularly with tribal leaders to pressure them to act.

One result of all these activities was a study on the Pine Ridge Reservation by the Centers for Disease Control. However, the report, issued in late 1981, concluded that no unusual health problems could be detected—a finding WARN greeted with skepticism. Officials set up a new community health program in 1982, but its focus was on smoking, alcohol, and health care rather than on environmental hazards. Authorities took no action to eliminate the hazards, and no water was shipped in, despite a drought that left certain areas without drinking water for two weeks. WARN is continuing legal and political efforts to force the government to provide the reservation with safe water.

Case History 3: Coalition for a United Elizabeth, Elizabeth, New Jersey

The city of Elizabeth, New Jersey, lies in the shadow of one of the nation's largest petrochemical complexes.[6] Its 200,000 residents have as neighbors oil refineries, chemical plants, and the web of highways and railways that are used to transport their products. A New York Times reporter described Elizabeth's air as having "an unnatural, sometimes pungent smell, occasionally accompanied by gray or black smoke."[7] The New Jersey Department of Environmental Protection pronounced Elizabeth's air "unhealthy" on 15 percent of the days in 1980.

The city has had more than its share of environmental accidents. In May 1979 a drum of acid ruptured at an abandoned Chemical Control Corporation warehouse, releasing caustic fumes that injured two police officers. In May 1979, and again in April 1980, an Exxon refinery spewed a cloud of aluminum silicate, a fine white powder that can contribute to lung disease, onto parts of Elizabeth. In July 1980 hydrogen sulfide fumes sickened at least fifteen people in nearby Port Elizabeth. In January 1981 a boiler malfunction at the Allied Chemical Corporation sent sulfuric acid into the air. The next day an Allied Chemical truck carrying sulfur trioxide ruptured a safety valve, sending seventy Elizabethans to the hospital for treatment of eye, ear, nose, and throat irritation. In August 1982 a tank truck spilled hydrochloric acid solution onto the New Jersey Turnpike near Elizabeth, sending more than thirty-five people to the hospital.

In the most spectacular of Elizabeth's accidents, a dump site owned by Chemical Control exploded on Earth Day, 1980. Forty-five thousand drums of illegally stored chemical waste burned for more than ten hours, forcing the hospitalization of sixty-six people. Only a chance change in wind direction averted the necessity of evacuating tens of thousands of people from Elizabeth and nearby Staten Island.

This string of disasters helped to make toxic chemicals a major political issue in Elizabeth. Since 1975 more than one hundred community groups have joined the Coalition for a United Elizabeth (CUE) in order to address city problems such as poor housing, racism, and inadequate schools. By 1978 residents had begun to complain to CUE about air pollution and toxic chemical dump sites. When a television journalist reported in early 1979 that the Chemical Control site posed a danger to city residents, CUE decided to act. Its members contacted local newspapers, elected officials, and the state health department, urging the state to clean up the site. It sponsored a public forum on chemical waste, where more than three hundred community residents questioned scientists and public officials about the dump site's dangers.

Public pressure eventually forced the state to take over the Chemical Control operation. The Department of Environmental Protection inspectors who visited the site in 1979 found rusty and leaking barrels, unmarked containers, and sloppy housekeeping. Tests showed that substances stored there included dioxin, polychlorinated biphenyls (PCBs), benzene (a cancer-causing solvent), and low-level radioactive waste.

At the time of the explosion and fire, state officials were still debating how to clean up the mess at Chemical Control. After the fire, CUE took several actions to assess the effects of the disaster and prevent its reoccurrence. It established a scientific advisory panel to analyze health problems in the community. The panel found that several of the fire fighters who had put out the blaze had continuing lung problems. CUE maintained public pressure on the Department of Environmental Protection to finish the cleanup at Chemical Control. It also lobbied for stronger state and federal legislation to fund the cleanup of toxic dump sites, and urged the local prosecutor to act against Chemical Control's owners. Subsequent court testimony revealed that the company's owners had ties with organized crime.

Finally, CUE joined regional networks such as the Greater Newark Bay Coalition Against Toxics, which opposed planned toxic waste facilities in nearby Newark. In 1983 it initiated a statewide coalition against the transportation of radioactive materials. According to Sister Jacinta Fernandes, CUE's associate director, this "networking helps people to become more aware in general, and it also gives more clout to the whole fight. We've had a lot of correspondence with other groups around the country who have wanted information from us and sent us their information. I think as people network around the country, it will put more pressure on the government, not only to clean up what they do have, but to prevent future problems."[8]

All across the country the air we breathe, the water we drink, the food we eat, the land on which we build our houses and schools, and the highways we travel are being contaminated by a constantly expanding brew of health-damaging substances. As a result, all across the country concerned citizens are joining together to fight the corporations, government agencies, utility companies, and agribusiness interests that are poisoning our families and communities.

How has this corporate and governmental assault on human health come to dominate the lives of so many people in the United States? Why have the residents of Love Canal, New York, Three Mile Island, Pennsylvania, Times Beach, Missouri, and thousands of other communities been forced to organize to protect themselves?

The roots of the current epidemic of environmental disasters go back to changes in U.S. industry during and after World War II. Scientific advances in organic chemistry, biochemistry, and other basic sciences provided the foundation for technological innovations. The war speeded the application of these new technologies to industrial production, and the most spectacular advances were in the petrochemical industry, which produces everything from plastics to pesticides. Between 1940 and 1980 the production of synthetic organic chemicals increased from less than 10 billion pounds a year to more than 350 billion.[9] Each year, one thousand new compounds are added to the estimated seventy thousand chemicals already in use by 1980.[10]

The growth of the petrochemical industry has been both a cause and an effect of other industrial development. The burgeoning automobile industry created a steady market for petroleum products. When many families became able to afford cars, developers were able to build new suburbs, new highways, and new industrial parks, all requiring a myriad of the new substances turned out by industry.

The petrochemical industry also revolutionized agriculture. During the past decades chemical fertilizers, pesticides, and preservatives have become an essential aspect of U.S. food production. There are now approximately 40,000 pesticide products on the market.[11] Between 1942 and 1967 chemical fertilizer use in the United States expanded tenfold.[12]

World War II served as the midwife to the nuclear industry. Massive government funding spurred the research necessary to build the atomic bombs eventually dropped on Hiroshima and Nagasaki, and the military has since continued to produce ever growing stockpiles of nuclear weapons. In the 1950s tax dollars also helped to build the nuclear power industry. By 1978 seventy nuclear power plants were generating 12.5 percent of the nation's electrical power.[13] Although economic problems and a militant antinuclear movement now threaten the nuclear power industry's growth, dangerous waste products will remain in existence for generations to come.

The petrochemical industry, agribusiness, military production, and the nuclear power industry all developed into major economic forces in the years after World War II. Their environmental impact is now added to that of more traditional manufacturing activities. Each stage of production contributes to the assault on health. The dangers begin with the extraction of raw materials from the earth. Mining of coal, lead, uranium, and other substances causes a wide range of health problems. Let us look at a few examples.

As the costs and risks of nuclear power increase and the price of oil goes up, coal has again become an important source of energy. Between 1975 and 1981, the amount of coal burned by utility companies increased by 50 percent.[14] Coal mining affects miners most directly. Between 1978 and 1980, 40,000 miners were killed or injured in accidents in underground coal mines alone.[15] Between 1970 and 1977 more than 400,000 miners or their widows were awarded compensation for black lung disease, a chronic condition caused by exposure to coal dust. Such exposure slowly damages its victims' ability to breathe, eventually killing them.[16]

Mining coal also affects the health of people in the communities near the mines. Coal dust can cause lung disease in children living near strip mines,[17] and mine fires increase air pollution, contributing to respiratory disease in nearby areas.[18] Many Appalachian communities have been unable to use their drinking water because it is contaminated by coal dust.[19] And coal mining can lead to explosions, flooding, and landslides. In 1973, in Buffalo Creek, West Virginia, a pile of coal waste collapsed after water had built up behind it. One hundred and twenty-five people were killed when the valley behind the waste site was flooded with water and coal debris.[20]

Lead mining can cause lead poisoning, a condition associated with learning problems and brain damage. In Kellog, Idaho, a 1974 government study found that 40 percent of the children living near a lead smelting plant had dangerously high levels of lead in their blood.[21] Smelters melt down raw ore after it is mined in order to extract the purified metals (in this case lead), and in the process discharge lead into the air. The study identified the Bunker Hill Company of Kellog as the source of the country's worst mass exposure of children to lead. Several years later, in 1979, tests showed that airborne lead levels in the Silver King Elementary School, just below the smelter's stacks, registered ten times the federal safety level.[22] Two families whose nine children had high blood lead levels in 1974 had their children tested by neurologists. When the doctors found all nine had nerve or brain impairment, the parents filed a $20 million lawsuit against the Bunker Hill Company. More recently, in 1983, lead emissions from a smelter in Dallas, Texas, led the director of the Dallas Housing Authority to recommend the relocation of four hundred families living in nearby public housing. Lead levels in a neighboring day care center had reached ninety-two times the federal standard.[23]

Uranium mining is also hazardous. Uranium miners, many of whom are Navajo Indians, are ten times more likely than other people to die of lung cancer. Dr. Victor Archer, a physician at the National Institute for Occupational Safety and Health (NIOSH), has estimated that one thousand of the six thousand miners who worked in one uranium mining area in the western Rocky Mountains will die of lung cancer.[24]

After raw uranium ore is removed from a mine, it is ground into a sand at a mill so that the uranium can be extracted. Each ton of ore yields no more than four pounds of quality uranium, or "yellowcake." The radioactive waste that remains, known as uranium tailings, is dumped near the mines and mills, and is often used as landfill. This radioactivity contaminates the environment, putting whole communities at risk. People living near a uranium mine in Colorado found high levels of radioactivity in their drinking water.[25] The Pine Ridge Indian Reservation in South Dakota, described at the beginning of this chapter, is downwind from deposits of uranium tailings from abandoned mines. Researchers suspect that high rates of miscarriage, certain kinds of birth defects, and cancers are related to the exposure to radiation through the water and air.[26] An EPA study found that people living within half a mile of uranium tailings are twice as likely to die of lung cancer as is the rest of the population.[27] Other problems associated with low-level radiation include premature death and leukemia.[28] Most experts agree that there is no safe level of exposure to radiation.[29] Uranium deposits in the United States are concentrated in the southwest, 80 percent on Native American reservations.[30] Recently, energy companies have also begun to explore for uranium in New Jersey, New York, Virginia, Wisconsin, and several other states. Intense public opposition has succeeded in instituting legislative bans or moratoria in New York, New Jersey, and Virginia.[31]

Another mining hazard is beryllium disease, an often fatal condition that affects beryllium miners and their families.[32] Used in fluorescent lights and atomic weapons, beryllium is mined in Pennsylvania. Still another hazard comes from phosphate mining, which has contaminated drinking water supplies in southwest Florida.[33] Finally, the extraction of iron, copper, and other metals leads to both environmental degradation and lung disease among miners.

Mining and processing ores create one set of environmental threats to health; subsequent stages of energy production are no less damaging. Fuel combustion by electric utilities contributes significantly to air pollution by producing particulates and sulfur and nitrous oxides. Each of these increases the risk of respiratory and heart diseases.[34] A study by the Brookhaven National Laboratory and Carnegie-Mellon University, for instance, estimated that each year twenty-one thousand people living east of the Mississippi River may die prematurely as a result of pollutants from coal- and oil-burning plants.[35] Burning coal is a particularly important source of air pollution. It is a major cause of "acid rain," a phenomenon that is killing fish and other animals in hundreds of lakes in the northeastern United States and Canada. Recently, new evidence suggests that acid rain can contaminate drinking water supplies, leading to human illness.[36] Coal combustion also releases carbon dioxide into the atmosphere, possibly causing long-term modification of the climate.[37]

Many of the new coal-based synthetic fuels pose similar hazards. Although synfuel processing removes many of coal's impurities, the wastes are then dumped, polluting the air and water with dangerous metals.[38]

Three Mile Island and the antinuclear movement put safety issues in nuclear power plants on the front pages of newspapers across the country. But the nuclear power industry's safety record since the Three Mile Island accident in 1978 is hardly reassuring. In 1980 the sixty-nine licensed and operable atomic reactors in the United States reported 3,804 mishaps, including spills, leaks, and equipment malfunctions.[39] The Nuclear Regulatory Commission's (NRC) first national survey, in 1981, found fifteen of fifty plants "below average" in management control, maintenance, radiation and fire protection, and overall compliance with operating regulations.[40]

While a power plant meltdown is the most serious danger, the daily release of small amounts of radioactivity into the air and water creates a continuing risk. The most persistent problem, however, is the disposal of high-level radioactive waste. Each year a single nuclear power plant generates about thirty metric tons of such waste,[41] some components of which can cause harmful health effects for up to 250,000 years.[42] Yet no permanent safe disposal site for this waste has been found. Instead, it is shipped around the country by truck, train, and barge. The Oak Ridge National Laboratory estimated that the nuclear industry will have to transport over 75,000 truckloads of used and dangerous nuclear fuel over the next twenty-five years.[43] The consequences of an accident could be disastrous. Dr. Leonard Solon, the director of the New York City Department of Health's Bureau of Radiation Control, testified before a 1978 Department of Transportation hearing in Washington that "dispersion of even a small fraction of these shipments as the result of an air crash, concomitant fire, and high winds within the City of New York . . . could have cataclysmic results, bringing death or serious injury to thousands of New Yorkers."[44]

Electricity generated by nuclear or coal-based plants is transmitted to the consumer. When electrical energy travels through extra-high-voltage power lines, it causes a variety of problems. Farmers whose animals graze under high-voltage power lines report that their cows have difficulty feeding their calves.[45] Laboratory experiments have demonstrated that mice exposed to electromagnetic fields similar to those produced by power lines lose weight and experience increased rates of infant mortality.[46] One study by researchers at the University of Colorado Medical Center found that children with cancer were more likely to live in homes close to high voltage electrical lines than were healthy children.[47]

Manufacturing chemicals, plastics, and other products poison industrial workers and pollute the air and water for everyone else. Because workers experience the most direct exposure, their toll is highest. According to NIOSH, each year 100,000 workers die from occupational illnesses, including lung cancer, heart disease, and respiratory conditions, and each year there are 400,000 new cases of work-related disease.[48] Nine out of every ten U.S. industrial workers are inadequately protected against exposure to toxic chemicals.[49]

But toxics do not stop at the factory gate. Through smoke stacks and drainage pipes, from storage tanks, and even carried by strong winds, dangerous substances escape from the workplace into the community, where they can contaminate the air, the water, and the soil. Industrial production accounts for about 30 percent of particulate air pollution, and adds sulfur dioxide, nitrogen oxide, hydrocarbons, and carbon monoxide to the air.[50] Although scientists debate the precise impact of air pollution on health, there is solid evidence linking industrial pollutants to such health problems as chronic bronchitis, asthma, emphysema, and other respiratory illnesses, as well as to heart disease and cancer.[51] Residents of communities with many chemical factories, such as Elizabeth, New Jersey, frequently complain of pollution-related eye, ear, nose, and throat irritation.[52] Foul odors from factory emissions, while not necessarily health-damaging, can certainly make life unpleasant.

In testimony before the House Subcommittee on Health and the Environment in 1977, Dr. Carl Shy, former director of research on air pollution for the federal government, estimated that failure to meet legal air quality standards may mean that an additional 3 percent of the population (or 7 million people) will suffer more severe asthma, an additional 10 to 15 percent of the exposed adult population will face a higher risk of chronic bronchitis and emphysema, and 100 percent of exposed children will face an increased risk of disturbed lung function.[53]

Epidemiological studies have found higher than average cancer rates in heavily industrialized areas. The industrial areas of California's Contra Costa County, for example—the home of five oil refineries and thirty-seven chemical plants—have a death rate from lung cancer for white males that is 40 percent above that of the rest of the county.[54] The death rates for blacks of both sexes and for white females are also elevated in these same areas. Since blacks and white women have generally been excluded from work in the chemical industry, these figures suggest that air pollution as well as workplace exposure is an important factor. In New Orleans, another center of petrochemical production, white males have a death rate from lung and bronchial cancer that is 40 percent above the national rate for this group.[55]

In New Jersey, a state with high cancer death rates and heavy industry, researchers have found significant levels of airborne pollutants such as benzene (known to cause leukemia), carbon tetrachloride (known to cause liver cancer), and the carcinogens arsenic, cadmium, chromium, and nickel in three industrial cities.[56]

Industrial production can also contribute to water pollution. Some manufacturers store raw materials in containers that leak into the groundwater. For example, in the city of San Jose, in California's "Silicon Valley," just south of San Francisco, mothers became alarmed at what seemed to be a large number of miscarriages and birth defects.[57] Tests revealed high levels of trichloroethylene (a substance suspected of causing cancer) and the toxic trichloroethane in the drinking water. State health department investigators discovered that several nearby manufacturers of silicon chips were storing these chemicals in underground tanks and that these were leaking. In 1983 an EPA official testified before the Toxic Substances Subcommittee of the Senate Committee on Environment and Public Works that between 75,000 and 100,000 storage tanks were leaking 11 million gallons of gasoline a year. He went on to say that one gallon a day leaking into an underground water source was enough to pollute the drinking water of a community of fifty thousand people.[58] Another important source of water pollution is improper disposal of toxic chemical waste, a problem that will be discussed in the section on waste disposal.

After a product has been mined or manufactured, it must be transported to its sellers and users. The growth of the petrochemical industry has led to a corresponding increase in the shipment of toxic chemicals and other hazardous materials. Every second, more than 125 tons of hazardous chemicals move onto the nation's highways, railways, and waterways.[59] Over four hundred thousand trucks regularly transport hazardous materials;[60] at least 4 billion tons are shipped each year. Moreover, the number of shipments of raw materials, finished products, and wastes is expected to double in the next decade.[61]

The U.S. Department of Transportation classifies 1,600 materials "hazardous." They range from poisons and explosives to gasoline and ammonia. In 1978 there were 18,000 accidents involving the rail, highway, or water transport of these substances, eight times the number in 1971. Injuries to drivers and passersby from these incidents increased by 400 percent in the same period.[62] In Waverly, Tennessee, in 1978, the explosion of a tank car carrying propane gas killed fifteen people and leveled the downtown section of the city.[63] In the same year a railroad accident that led to a spill of the toxic chemical phosphorous trichloride in Somerville, Massachusetts, forced the evacuation of thirteen thousand people and sent 200 to the hospital.[64] In Montebello, California, in 1981, thousands of people became nauseated and dizzy when a truck leaked ethyl mercaptan, a foul-smelling toxic substance, for twenty-five miles along a freeway.[65] And in 1982 a train derailed near Livingston, Louisiana, spilling more than 1 million pounds of such toxic chemicals as vinyl chloride, toluene, phosphoric acid, and tetraethyl lead. The ensuing explosion set off a fire that burned for two weeks, forcing authorities to evacuate 2,700 people.[66] When residents were allowed to return home two weeks later, the health department warned them not to eat vegetables from their gardens or food that had been left on shelves or even in refrigerators.

After a product has been delivered to its consumers, it can endanger health in new ways. Asbestos, a substance used for insulation and fireproofing, has become a pervasive hazard not only to its producers but also to its users. Exposure to flaking asbestos can permanently damage lungs, and it is now known that even brief exposure can lead to cancer.[67] Hundreds of thousands of schools and other buildings around the country have used asbestos tiles or insulation. According to surveys by the EPA and by the union representing school janitors, 14,000 of the nation's 110,000 schools have an asbestos problem, posing a cancer risk to 4 million students and school employees.[68] Since as much as twenty years can elapse between exposure and the first sign of cancer, millions of Americans are condemned to decades of fear and uncertainty.

At high temperatures, polyvinyl chloride (PVC), a plastic used in floor coverings, phonograph records, tubing, insulation, and food wrappings, can release fumes that cause liver and other cancers, as well as kidney damage. Recently, when the New York City Transit Authority installed PVC tubing to protect the public address system in several subway stations, scientists charged that subway fires, frequent in the New York City system, could melt the tubes, releasing fumes that could kill or injure thousands of subway riders.[69] The tubing was eventually removed.

Another substance used for insulation, polychlorinated biphenyls, or PCBs, can cause cancer and skin disease. Exposure to PCBs has also led to birth defects in laboratory animals. In 1981 an electrical transformer in a state office building in Binghamton, New York, exploded during a fire, showering the eighteen-story building with PCBs and their by-products, including the deadly poison dioxin. The building had to be abandoned. A cleanup would cost more than $20 million—for a structure that cost only $13 million to build.[70] Several people who worked in the building suffered liver damage as a result of PCB exposure. In 1983 there were 1.8 million capacitators and 100,000 transformers containing PCBs in the United States. According to Dr. Arnold Schecter, a professor of preventive medicine at the Upstate Medical Center in Binghamton, "PCB transformers are public health time bombs, medical time bombs, waiting to go off."[71]

Some of the products used to control termites have forced owners to abandon their homes. After an exterminator sprayed aldrin, a banned cancer-causing pesticide, in a Long Island house, the New York State Department of Environmental Conservation advised the family to vacate the premises within two days.[72] When laboratory tests confirmed high levels of the persistent chemical, the owner demolished his house and buried the rubble in a landfill. A subsequent investigation by state authorities found widespread use of aldrin, as well as heptachlor and chlordane, two other carcinogens used in termite control.

Other toxic substances that can harm consumers include formaldehyde, a product widely used in home insulation and textiles and known to cause cancer in laboratory animals;[73] building materials contaminated with radiation from uranium tailings; and the thousands of toxic substances used daily by homemakers and hobbyists.[74]

The forced evacuation of more than 800 families from their homes at Love Canal in Niagara Falls, New York, in 1980 alerted the public to the growing problems caused by the improper disposal of chemical waste. In 1981 more than 50,000 companies in the United States produced 165 million tons of hazardous waste, enough to fill 7,500 Love Canals.[75] The chemical industry was responsible for more than 70 percent of these. Exposure to toxic waste can lead to eye, ear, nose, throat, and skin irritations, breathing problems, kidney and liver damage, miscarriages, birth defects, and cancer.[76] People come into contact with toxic chemicals most commonly by drinking water contaminated by improperly sealed dump sites. The case of Hardeman County is only one of many such situations. Direct exposure to such waste is also common. For example, children playing near a chemical dump site in Hammond, Indiana, developed skin rashes,[77] and seepage of waste into the basements of the homes at Love Canal led to dangerous levels of chemicals inside the houses.[78] A report issued by the Congressional Office of Technology Assessment in 1983 estimates that to adequately protect public health, it will cost as much as $40 billion to clean up the fifteen thousand sites where toxic waste is known to have been dumped without controls.[79] A 1982 EPA study concludes that more than 90 percent of the eighty thousand sites where toxic chemicals are stored in pits, ponds, or lagoons threaten to contaminate the ground water.[80]

The conditions at hazardous waste disposal facilities vary widely. The best operators bury containers of chemicals in clay-lined pits with carefully planned drainage systems, and then carefully monitor the runoff. But William Sanjour, chief of EPA's hazardous waste implementation branch, notes that all landfills, even those with an "impermeable" clay cap, eventually leak. "So," he told the *New York Times*, "the federal government cannot come out with any set of regulations that protects human health and the environment and allows existing landfills to continue to function."[81] As we will see in chapter 3, many corporations ignore federal and state regulations altogether, simply paying a carter to dump their wastes in an abandoned lot, an empty warehouse, or a wooded area.

In the last five years, more and more communities built near toxic dump sites have experienced serious health problems. Although investigators have not always been able to identify the specific cause of illness, the anecdotal evidence leaves many scientists alarmed. Dr. Philip Landrigan of NIOSH has noted that "there are already sufficient quantities of hazardous wastes in our environment to provide a legacy of disease and death to our descendants for generations to come."[82] Dr. Samuel Epstein, professor of environmental medicine at the University of Illinois School of Public Health, has called improper disposal of hazardous wastes "America's number one environmental problem."[83] A few reports from affected areas illustrate the dimensions of the problem.

In Woburn, Massachusetts, chemical companies, tanneries, and a glue-making factory dumped their wastes into empty lots for more than a century, contaminating a sixty-acre area with arsenic, lead, and chromium. Federal officials have listed Woburn as one of the ten worst hazardous sites in the country.[84] This city of thirty-five thousand people has one of the highest cancer rates in Massachusetts. Between 1969 and 1978 its rate of childhood leukemia was more than double the national average and the rate of kidney cancer was also unusually high.[85]

In Farmington, New Jersey, a small community near Atlantic City, residents complained of skin rashes and a large number of cancer deaths. For years a nearby landfill had been used by corporations such as DuPont, Union Carbide, and Procter and Gamble to dispose of toxic chemical wastes.[86] Tests of Farmington's drinking water in 1981 showed staggering concentrations of dangerous chemicals. Levels of benzene, a cause of leukemia and of aplastic anemia, were 11,000 times higher than "safe" levels. Another chemical, dichloroethane, which causes kidney and liver damage, was detected in concentrations 136,000 times greater than the EPA safety level. Other dangerous substances found in the water included arsenic, vinyl chloride, lead, mercury, cadmium, and chloroform. Farmington's inhabitants now drink bottled water. Even worse, the dump that ruined their water threatens to contaminate Atlantic City's water supply. State health officials have closed many of the wells that supply water to the resort's forty thousand year-round residents and to the millions of tourists, and are still searching for alternative sources.[87]

Recently, waste disposal firms have found new and ingenious methods to expose people to toxics. In 1982 officials from the New York City Department of Environmental Protection found illegally high levels of sulfur, benzene, and lead in fuel tanks in residential apartment buildings.[88] According to state investigators and industry spokespersons, fuel retailers and wholesalers sometimes mix "cocktails" of toxic waste and heating oil in order to boost profits or undercut competitors.

The toasters, automobiles, cosmetics, and frisbees produced by U.S. manufacturers have left the people of this country a legacy of illness and death. And as big business has taken over food and forestry production, agriculture has also contributed to poor health, primarily by its increased use of petrochemical products. In the last two decades, for example, the production of synthetic organic pesticides in the United States has doubled.[89] The use of herbicides, which rid land of unwanted vegetation, has grown even more dramatically. In 1977 250 million acres of land were sprayed with herbicides, as compared with 71 million acres in 1962.[90] In 1982 alone, herbicide sales increased by 20 percent.[91]

More than twenty years ago, Rachel Carson's *Silent Spring* alerted the public to the potential dangers of pesticides. In a "fable for tomorrow," she described a community where

> mysterious maladies swept the flocks of chickens; the cattle and sheep sickened and died. The farmers spoke of much illness among their families. In the town the doctors had become more and more puzzled by new kinds of sickness appearing among their patients. . . . Even the streams were now lifeless. Anglers no longer visited them, for all the fish had died.[92]

Two decades later, more and more communities experience the reality of Carson's grim prophecy.

As we would expect, those who make pesticides and those who use them on the job—farm workers, forestry workers, and those who spray rights-of-way for the utilities and railroads—suffer the most serious consequences,[93] and each year 45,000 people (mostly workers) are poisoned.[94] Their symptoms include nausea, skin rashes, neurological disorders, and respiratory problems. Exposure to insecticides (pesticides used to kill insects) such as dieldrin, kepone, endrin, lindane, mirex, heptachlor, and toxaphere can cause cancer in humans or animals.[95] Studies have also linked certain pesticides with sterility.[96]

Those who live near land used for forestry or farming are also exposed. Communities in the Pacific Northwest, for example, have reported a high incidence of miscarriage and birth defects after herbicides were sprayed on commercial and government-owned forests.[97] In Scottsdale, Arizona, an upper-middle-class community bordered by farmland, crop dusters fly over the cotton fields, spraying them with pesticides. In 1978, 343 residents complained to the Arizona Pesticides Control Board that the drifting spray had made them sick.[98]

Pesticides can enter the human body through air, water, or food. The increasing use of substances such as carbaryl (to control gypsy moth caterpillars) and malathion (to control the infamous Medfly in California) in urban and suburban areas means that more and more peope are directly exposed to airborne pesticides. Overall, 65 percent of pesticides are sprayed by plane, which greatly increases the number of people exposed by wind drift.[99]

As substances sprayed on crops or forests drain into underground waterways, rivers, or streams, drinking water is contaminated. Pesticides have been found in drinking water throughout the country. In rural Suffolk County on Long Island, for instance, the pesticide Temik, used to control potato bugs, was found in several wells in 1979. A year later public outrage forced Union Carbide, the manufacturer of Temik, to install water filters in 1,700 homes whose water was contaminated,[100] and EPA banned its sale in Suffolk County. A preliminary study by the Suffolk County Health Department in 1982 showed that pregnant women living in homes with Temik-contaminated water had two times more miscarriages than did women whose water was less contaminated.[101] In 1983 Cornell University scientists predicted that Suffolk's water might remain polluted with Temik for more than 100 years.[102] Ground water in Wisconsin and Florida, where

Temik is still legally sold, has recently been discovered to contain high levels of the pesticide.

In California's agricultural Fresno County, fruit growers killed worms that attacked the roots of trees and vines with the pesticide DBCP until it was banned in 1977 after workers making the product became sterile. But five years later, according to the California Health Department, 12,000 people in Fresno County and as many as fifty thousand in the San Joaquin Valley were still drinking water with unsafe levels of DBCP, which had contaminated public water supplies.[103] Even more alarming, another health department study found that people living in areas where the water was contaminated with DBCP had higher rates of stomach cancer than did the rest of the population.[104]

Pesticides also enter our bodies on the food we eat. Almost all commercially grown fruits and vegetables are sprayed at some stage of their production. An incident in Hawaii in 1982 illustrates one way that such contamination occurs. After the state health department found that milk on the island of Oahu contained from three to six times the accepted level of heptachlor, a carcinogenic pesticide that causes liver and kidney damage, it recalled all milk products from grocery shelves and school cafeterias.[105] Investigators discovered that dairy cattle had been fed with pineapple leaves sprayed with heptachlor. A 1980 EPA study estimated that 93 percent of the population has in their bodies some levels of dieldrin, a carcinogenic pesticide that was banned in 1976.[106] Pesticides are also found in meat and dairy products. According to the Food and Drug Administration, between 500 and 600 toxic chemicals may be present in this country's meat supply.[107] Carol Tucker Foreman, assistant secretary of agriculture from 1977 to 1980, testified before Congress in 1982 that "there is a good chance that the American public consumes meat with violative levels of carcinogenic and teratogenic [causing birth defects] chemical residues with some regularity."[108]

Food can also become contaminated by other substances. In 1973 in Michigan, 1,000 to 2,000 pounds of PBBs, a flame-retardant, were accidentally substituted for a feed additive for farm animals. PBBs are known to damage the nervous system, the liver, and the immune system. As a result of the accident, 30,000 cattle and 1.5 million chickens had to be destroyed.[109] More ominously, five years after the incident a research team from Mt. Sinai

Hospital in New York City estimated that 97 percent of Michigan residents had some level of PBBs in their bodies.[110]

Modern agricultural production also contaminates drinking water with fertilizers, processed foods with additives and preservatives, and meat with hormones and drugs. Each of these forms of contamination can lead to illness, including cancer, kidney and liver disease, and allergic reactions.[111]

The U.S. military uses many of the same chemicals and production processes as does private industry, and its impact on the environment requires close scrutiny by environmentalists. Most dramatically, nuclear weapons threaten the ultimate environmental disaster: the total destruction of life and civilization. In economic terms, the Department of Defense controls vast resources. Its 1984 budget was $250 billion. It operates more than four thousand military installations in the United States alone; more than 80 private companies work in its nuclear weapons production program.[112] Moreover, the military is involved in producing, storing, testing, and transporting a variety of lethal agents.

Like other industries, the military threatens its own workers most directly. At a press conference in Sacramento, California, in 1982, a former army medic testified that he had been ordered to report false data to cover up the fact that soldiers at atomic tests in the 1950s were exposed to dangerously high levels of radiation.[113] A 1982 study by the U.S. Centers for Disease Control found that the 3,027 men who took part in one of the Nevada tests had leukemia death rates three times higher than the population at large.[114] Since the end of the Vietnam war, U.S. veterans have complained of rashes, birth defects in their children, and nervous conditions that they attribute to their exposure to Agent Orange, a herbicide that was used to defoliate jungle areas during the war. While U.S. scientists are only now beginning to study the long-term health effects of Agent Orange, Vietnamese researchers reported at a 1983 international conference in Ho Chi Minh City that they had found an increase in abnormalities in children of women whose husbands had been exposed to wartime use of the herbicide.[115]

Once again, however, the dangers of military programs do not stop at the barracks gate. In 1957 a fire at the Rocky Flats nuclear weapons plant in Colorado sent a radioactive cloud across nearby Denver. Years later Dow Chemical Company, which operated the plant for the government, found high levels of carcinogenic radioactive materials at two elementary schools within twelve miles of the Rocky Flats plant.[116] Dr. Carl Johnson, former director of the Jefferson County (Colorado) Health Department, called the Rocky Flats release in 1957 "the most important exposure to the population near the plant during the period 1953–71."[117] In 1981 the U.S. government agreed to an out-of-court settlement with a resident of Igloo, South Dakota, who contended that he and others in the area had been poisoned by the destruction of toxic gases at the army's Black Hills Ordnance Depot.[118] According to army documents obtained by South Dakota's *Rapid City Journal* under the Freedom of Information Act, for at least twenty years the army had routinely burned and released into the air mustard gas, phosgene, and cyanogen chloride. The rancher who won the settlement had suffered severe lung damage, a known effect of gas poisoning.

In 1982, when residents of Seneca County in upstate New York learned that the Seneca Army Depot was the site of more nuclear warheads than any other base in the United States, many became worried. "This area would be a prime target if there was a war," said one woman. "But even if there isn't, all kinds of accidents could happen so that radioactivity could be released in the area and we'd never know about it."[119]

The military is a major generator of radioactive and toxic chemical waste, which poses major threats to the health of U.S. citizens. In New Brighton, a suburb of Minneapolis, for example, the Twin Cities Army Ammunition Plant has contaminated the 23,500 residents' supply of ground water with the carcinogen trichloroethylene.[120] Unfortunately, most environmental regulations exempt military programs; government secrecy further limits our knowledge of the magnitude of the problem. In the name of national security, the defense establishment has become nearly immune to public scrutiny and criticism, which makes corrective action doubly difficult.

Causes of the Current Crisis

These vignettes from across the United States indicate that every community in the country faces an actual or potential threat to its health. From pesticide spraying to uranium mining, from toxic dump sites to industrial air pollution, from nuclear power plants to the national network of highways and railways used to transport toxic chemicals, Americans are experiencing a chemical and radioactive assault on their health. New technologies that could have been used to feed, clothe, house, and transport people are instead destroying the environment that sustains human life and damaging the health of present and future generations. Why is this enormous misappropriation of our nation's resources allowed to take place?

The answer requires a closer look at our economic and political system. In the economic sphere, the pursuit of profit is the force that drives a capitalist system. Capitalists invest money in those industries that will return the highest rate of profit. They design production processes that will minimize their costs and maximize their returns. Within a corporation, directors decide how to allocate resources based on the need for profits.

Using profit as the primary mechanism for economic planning has many environmental consequences. It determines which technologies flourish and which die. In *The Closing Circle*, environmental scientist Barry Commoner describes how natural organic products such as soap and cotton were replaced after World War II with unnatural synthetic ones such as detergents and plastics.[121] In almost every case the new technology damaged the environment more than the old one, but in each case the new production process was also more profitable. In fact, the very qualities that made the new products profitable also made them destructive to the environment. Many plastic goods, for example, are used once and then disposed of, guaranteeing both a steady market for new products and growing piles of nondegradable, often toxic, garbage. Commoner explains the economic imperative:

> The quandary of the petrochemical industry is that the unique reasons for its success and growth—that it can produce a growing variety of new man-made substances, and can sell them cheaply, but only in very large amounts—are themselves the sources of its growing threat to society. Pressed by its economic structure to create ever more chemically complex man-made products on huge scales, the industry now confronts a hard fact of nature—that the more complex these products, the more likely they are to harm living things, including people, and the more widespread they are, the greater their toxic impact.[122]

The automobile industry provides another example of choosing profits over a cleaner environment. The average 1968 passenger car emitted more than twice the nitrogen oxide exhaust as the 1946 car. And cars in 1971 spewed out nearly twice as much lead per mile as models made in the postwar years. The major reason that automakers preferred powerful gas-guzzling cars was profit. In the late 1960s it cost General Motors only $300 more to make a Cadillac Coupe de Ville than the smaller Chevrolet Caprice. But consumers paid $3,800 more for the Caddy—the difference went into GMs coffers.[123] By 1983, as memories of the energy crisis receded, autombobile manufacturers were again persuading people to buy big, polluting—and profitable—cars.

Agribusiness's hunger for profits has led it to aggressively market products that can in the long run ruin farmland. For example, pesticides and fertilizers are like heroin: the more they are used, the more they are needed. Pesticides kill not only the pest but also its predators. As birds, larger insects, or frogs that eat the pest are killed off, the pest reproduces more rapidly. More pesticides are required to maintain production. Species-specific control agents are available, but broad-spectrum pesticides—those that kill many species—are marketed more widely because they are more profitable. Similarly, as soil becomes depleted by a single crop, such as wheat or corn, that only chemical fertilizers can sustain, increasing amounts of fertilizer are needed to maintain the same level of productivity. The farmers' costs go up and fertilizer runoff pollutes the water, but agribusinesses make healthy profits.

A desire for a fast return on investment also leads to more rapid extraction of natural resources from the earth, even though a slower pace would cause less damage to both the environment and human health. For example, better protection of uranium miners against radiation and proper disposal of waste products mean fewer casualties and less contamination of surrounding communities, but such health measures lower profits.

Corporate insistence on dividends for its shareholders, combined with lax regulation, leads to inadequate expenditures on pollution control. As a result, toxic wastes are often disposed of in cheaper, more dangerous, ways rather than safer, more expensive, ones. According to one chemical industry spokesman, an illegal waste disposal company charges $1 a barrel to haul away and dump toxic wastes, while safe, legal disposal can cost as much as

$45 a barrel.[124] Not surprisingly, most of the major chemical companies have been caught using illicit disposal firms. The continual spills and breakdowns at nuclear power plants further illustrate industry's reluctance to spend money on safety. After all, every dollar not spent on environmental protection is a dollar of profit.

Since government control of pollution is weak, the corporations do not have to pay for the consequences of environmental contamination. The real costs are borne by the taxpayers, consumers, and workers who pay higher taxes for cleanup efforts, travel further or move to find clean air or water, and face devastating medical bills. To use the dramatic example of Love Canal, Hooker Chemical spent $1.7 million to dispose of its waste in an abandoned canal that the company then sold to the Niagara Falls school board. Nearly twenty-five years later, when the Love Canal residents finally forced the state and federal governments to clean up the site, New York State taxpayers alone paid out more than $61 million to remedy the situation.[125]

Skimping on environmental protection is one strategy for boosting profits. Another is to introduce new products quickly. A publication by the Manufacturing Chemists Association explains that "the maintenance of above average profit margins requires the continuous discovery of new products and specialties on which high profit margins can be earned."[126] By capturing the market before a competitor steps in, corporations can ensure several years of good returns. As a result, manufacturers introduce hundreds of new pesticides and other chemical products into the market each year. But the rapid journey from laboratory to marketplace means that there is no time for realistic evaluation of the environmental or health impact of a new product. The disasters caused by such toxins as PCBs, dioxin, PVCs, asbestos, and formaldehyde could have been avoided by proper testing, but a company that tested its products thoroughly would have had to sacrifice first crack at the market, thus violating its obligations to its shareholders to bring in steady profits.

In sum, the U.S. economy creates incentives for environmental degradation. It encourages capitalists to find the cheapest, quickest way to make profits, no matter what the environmental cost. When laws allow corporations to pass the costs of pollution on to the public rather than their shareholders, these corporations are further encouraged to pollute. And as we will see in chapter 3, industry uses its power and wealth to ensure that government regulations will not seriously interfere with making a profit.

The causes of the corporate assault on health are political as well as economic. The corporations that dominate the U.S. economy also shape our political system. Not only do they insist on their right to make a profit as they see fit, but they also want the power to construct a society that they can continue to control without undue public interference. In Hardeman County, Tennessee; Pine Ridge, South Dakota; and Elizabeth, New Jersey, corporate decisions endangered the health and property of the communities' residents. Yet these people never decided to expose themselves to chemical or radioactive hazards, just as the people of Love Canal never voted to allow Hooker Chemical to poison their neighborhood.

In the United States, citizens do not have the right to control what goes on in their communities. Rather, democracy has come to mean a relatively narrow freedom of choice: the right to choose a candidate for office, to read whatever newspaper or magazine we like, to select a favorite brand of cigarette or designer jeans. Decisions about where to build or move factories, where to dump toxic waste, or what technologies to invest in are made by corporate boards and the managers they hire. When community residents or workers contest these decisions, they must file expensive lawsuits, lobby distant legislators, or pressure reluctant bureaucrats, all in an arena in which corporations help to set the rules and choose the players.

To maintain its power, industry also seeks to define the nation's political agenda. Its advertising, its advocates in Congress and the White House, and its academic supporters insist that only continued economic growth controlled by industry can ensure (or return) prosperity. The only legitimate question for these people is how to encourage such growth. Those who challenge continued corporate hegemony meet stiff resistance. Workers who refuse to sacrifice their health to increased profits face environmental blackmail—the threat of losing their jobs. Those who question the benefits of nuclear power or pesticides are labeled as extremists, impractical romantics, or communists.

But the continuing toll that industry's growth has taken on the health of the people of the United States has led increasing numbers of them to question a political and economic system that puts profit before human need. Across the country, community groups are fighting for a deeper concept of democracy, one that gives people a voice in making decisions about their neighborhoods, their communities, and their workplaces, and one that gives them the information necessary to make these judgments. At the root of this new movement lies the belief that people have the right to air, water, food, and land that will not sicken them, and the determination to create a society that guarantees this right.

2. SCIENCE AND POLITICS:
The Limitations
of Environmental Health Research

Case History: Love Canal, New York

In 1953 the Hooker Chemical Company, a subsidiary of Occidental Petroleum, deeded to the Niagara Falls Board of Education a plot of land it no longer needed.[1] For $1 the Board of Education accepted land that Hooker acknowledged had "been filled, in whole or in part, to the present ground level with waste products resulting from the manufacturing of chemicals." The deed also stated that the board "assumes all risks and liabilities incident to the use thereof and no claim, suit, action or demand of any nature whatsoever shall be made by the [board] . . . for injury to a person or persons, including death resulting therefrom, or loss of or damage to property caused by reason of the presence of said industrial wastes."[2] By 1955 four hundred children attended the new elementary school constructed on top of the chemical-filled Love Canal. In 1958 residents of the expanding community began to complain to City Hall of explosions at the canal site, of children being burned by chemicals, of nauseating odors, and of black sludge oozing from the ground. However, their persistent requests for a city investigation were either ignored or denied. Not until 1976, after a few local reporters wrote about the Love Canal residents' complaints of odors and rashes, did the New York State Department of Environmental Conservation test for toxic chemicals. A year later the U.S. Environmental Protection Agency (EPA) hired a consultant to collect air samples from the basements of a few Love Canal homes and also commissioned studies of the ground water. Eventually over two hundred different chemical compounds were identified in or around the canal, including twelve known cancer-causing substances.

In August 1978 Dr. Robert Whalen, New York State Commissioner of Health, called a press conference to reveal the state's plans for remedial action at the dump site. He reported that a state health department study had shown "a slight increase in risk for spontaneous abortions among all residents of the canal," and concluded that the "Love Canal Chemical Waste Landfill constitutes a public nuisance and an extremely serious threat and danger to the health, safety, and welfare of those using it, living near it, or exposed to the conditions emanating from it."[3] He advised that "as soon as possible" those pregnant women living nearest the canal "temporarily move from their homes" and that families in the area "relocate temporarily any children under two years of age."[4]

Residents were outraged at the commissioner's advice. "You're treating us like the Titanic! Women and children first," exclaimed Lois Gibbs, the leader of the Love Canal Homeowners' Association, a group she had helped organize earlier in 1978.[5] If the canal was dangerous for pregnant women and young children, people reasoned, then it could hardly be safe for anyone else. Furthermore, they were enraged that the state had made no provision for paying the expenses of any evacuation. Why should they have to pay for Hooker's action and the government's inaction?

Whalen's evacuation recommendation and the media attention generated new scientific research. One project was carried out by the Love Canal residents themselves. With the help of Dr. Beverly Paigen, a cancer researcher at the Roswell Park Memorial Institute in nearby Buffalo, volunteers were trained as interviewers and began a systematic telephone survey of families in the area. The survey found higher than average rates of miscarriage, kidney and bladder disorders, and central nervous system disorders among the respondents—and more than 75 percent of the families participated.[6] The residents believed that these findings supported their contention that the neighborhood was dangerous to their health.

Department of Health officials found the community's findings interesting, but since the research could not tie a specific chemical to a specific health effect, they tried to discredit the survey. A Department of Health physician called the data "totally, absolutely and emphatically incorrect"; another said it was "information collected by housewives that is useless."[7] Then in February 1979 the Department of Health issued a set of recommendations based on its own research. The state found no serious health problems among the children and adults living near the canal, but it did find higher rates of miscarriage, birth defects, and babies with low birth weights. As a result, the area that pregnant women and young children were advised to evacuate was expanded. Residents who had previously been told their homes were safe were furious. "My child was under two last August. It's your fault we didn't leave then," one parent shouted at the new health commissioner, Dr. David Axelrod.[8] Increasingly, the residents were coming to believe that state officials made policy decisions based not on scientific criteria but on political factors. This belief contributed to their mistrust of the scientists who were hired by the same body—the state government—that might have to pay for evacuation and cleanup.

The political climate in which scientific research took place was especially clear to Dr. Beverly Paigen, the researcher who assisted the Homeowners' Association. After her public testimony in support of the residents in Albany and Washington, she was notified that the Commissioner of Health had refused to allow her to receive a large government grant that she had applied for— even though it was for research on another topic.[9] She was also asked by her superiors at Roswell Park to give them advance copies of any speeches or papers she planned to deliver in public.[10] Finally, in August 1979 she was ordered to report for an audit of her New York State income tax. When she met the auditor, she noticed that her file was filled with news clippings about her work at Love Canal. A year later the state Commissioner of Taxation apologized to Paigen for "errors in procedure which have inconvenienced and concerned you."[11]

In May 1980 another scientific controversy erupted over the health consequences of the Love Canal site. Dr. Dante Picciano, a respected cytogeneticist (one who studies the relationship between genetic abnormalities and disease), released the results of a pilot study that had been commissioned by the EPA to look at the genetic material of Love Canal residents. The study showed that twelve of the thirty-six residents examined had "increased frequencies of cells with chromosome breaks . . . [and] are at an increased risk of neoplastic disease [cancer], of having spontaneous abortions, and of having children with birth defects."[12]

Two days after being informed of the results of the genetic survey, angry citizens held two EPA officials "hostage" for several hours in the Homeowners' Association's office. Forty-eight hours after the hostages were released, President Carter declared a state of emergency at Love Canal, permitting the federal government to relocate—at its expense—seven hundred families living close to the canal. As Lois Gibbs noted, the "hostage" action resulted in their getting "more attention [from the White House] in half a day than we've gotten in two years."[13] In fact, it took many more months of community action and political maneuvering to implement the president's decision, but it nevertheless marked a significant victory for Love Canal residents.

Meanwhile, however, the EPA had proceeded to disavow Picciano's study. A review panel chaired by Dr. David Rall, director of the National Institute of Environmental Health Sciences, criticized the study for its lack of a control group and the small size of the sample, and declared that it provided an "inadequate basis for any scientific or medical inferences from the data (even of a preliminary nature) concerning exposure to mutagenic substances of residents in the Love Canal area."[14] Picciano himself had acknowledged the shortcomings of the study even before beginning it, but the EPA had refused to provide funding for a more complete survey.

The relocation of most of the Love Canal residents did not end the battle over the implications of the scientific research. In October 1980 the Panel to Review Scientific Studies and the Development of Policy on Problems Resulting from Hazardous Wastes, appointed by New York State Governor Hugh Carey and chaired by eminent physician Lewis Thomas, chancellor of Memorial Sloan-Kettering Hospital in New York City, issued a report that concluded that, "there has been no demonstration of acute health effects linked to exposure to hazardous wastes of the Love Canal Site . . . [and] that chronic effects of hazardous waste exposure . . . have neither been established or ruled out yet, in a scientifically rigorous manner."[15]

The Thomas Panel, as it was called, made the decision—highly unusual in the scientific community—not to provide other investigators with the data (much of it unpublished) on which its decision was based or to comment on or respond to criticisms of its report. Dr. Adeline Levine, a sociologist who has studied the Love Canal story, has noted that four of the five physicians on the Thomas Panel were directors of institutions dependent on the New York State Department of Health for their reimbursement rates, while senior officials of the department, including Commissioner Axelrod, attended all the panel's meetings—which were otherwise closed to observers.[16] It is hardly surprising, then, that the report was greeted with skepticism by many, including Love Canal residents. As one put it, "Gee! It's the miracle of Love Canal! You live on top of a couple hundred chemicals for twenty-five years and nothing's wrong! . . . They should really send all the world's waste here, because it's the only place on earth where such a miracle could happen."[17]

The state and federal government have continued to maintain this point of view. In July 1982 the EPA issued a report on its study, initiated in May 1980, that declared that the neighborhood around the Love Canal dump site was safe for resettlement.[18] The new report was based on an analysis of air, water, and soil samples and on a review of the health data. But it raised as many questions as it answered. In the one-month period prior to the report's publication, the U.S. Department of Health and Human Services' (DHHS) own panel of six experts voted to endorse the EPA's recommendations, reversed itself, and then, on the day before the report was issued, reaffirmed its initial conclusions.[19] Five of the eleven scientists asked by DHHS to review the health data in the report disagreed with the EPA's conclusion that the neighborhood was safe. One criticism was that the EPA used other neighborhoods in Niagara Falls as controls, yet its own data showed that Niagara Falls as a whole had higher than average levels of pesticides, organic solvents, and heavy metals in its soil and water. Dr. Ellen Silbergeld, a toxicologist with the Environmental Defense Fund, an advocacy group of lawyers and scientists, called the study a failure and suggested that the federal government withdraw it.[20] The Love Canal Homeowners' Association said the study resulted in "questionable data, unsubstantiated conclusions, and an immense waste of taxpayers' money."[21]

By early 1983, 182 families were living in the resettled area and 270 more were on a waiting list to purchase homes there. In June the Office of Technology Assessment (OTA), a bipartisan arm of Congress whose judgments are respected in the scientific community, issued a scathing report on the 1982 Love Canal study, finding its sampling techniques, choice of control group, statistical analyses, and laboratory testing deficient. The OTA concluded:

> With available information, it is not possible to demonstrate with certainty either that unsafe levels of contamination do or do not exist. There remains a need to demonstrate more equivocally that [the area] is safe now and over the long term for human habitation. If that cannot be done, it may be prudent to accept the original presumption that the area is not habitable.[22]

A few months after the OTA's report was released, EPA investigators found that a "significant migration of chemicals" had taken place into the resettled area, leading it to postpone a decision as to whether to sell more homes for at least two years.[23]

At Love Canal scientists played a critical role in the investigation and the decision as to whether—and how—to clean up the toxic chemical waste dumped two decades earlier by the Hooker Chemical Company. They sampled the air, water, and soil for dangerous substances; they conducted health surveys among the residents and tested their blood, urine, and chromosomes; they reported their findings to the press and public; and they advised the Homeowners' Association and the state and federal governments on how best to correct the situation. Community residents, industry, and government cited the scientists—often to conflicting ends—to justify virtually every action they took.

The Love Canal disaster is thus a good illustration of both the differing uses scientific data can be put to and the limitations of science as a tool for solving environmental health problems. Five years after the story made headlines, scientists still disagree about the effects that the chemicals have had on the residents' health. Every scientific finding became the subject of a controversy. Results that supported the residents' claims that their health had been seriously affected were rejected by the state and federal governments and by industry. The residents, on the other hand, refused to believe government studies that claimed to demonstrate that health problems did not exist. While such disagreements might lead us to believe that science is simply irrelevant, that is not the lesson to be learned. Rather, we need to understand the limitations of scientific research so that we can evaluate it for ourselves, then decide when and how to use it to our advantage.

This chapter therefore begins with a description of the methods scientists use to assess environmental health hazards. By understanding the strengths and constraints of various methodologies, we can generalize more effectively from research papers to our own communities. We then look at the way in which the social, political, and economic context of scientific research influences its priorities and outcomes. Environmentalists who grasp the politics of science are less likely to be disillusioned by its practice. Finally, we discuss the differing perspectives of scientists and citizens and how these differences shape the interaction between them.

How Health Hazards Are Assessed

Scientists collect data on the health consequences of toxic chemicals and other hazardous materials by exposing various living organisms to the substance under investigation and observing their reactions. In general, the simpler the organism, the quicker and cheaper the test. The most frequently used short-term test is called the Ames test. Bacteria are exposed to a substance (or substances) suspected of causing cancer, and their genetic materials are then examined. Since carcinogenesis—the process that leads to the formation of a cancer—is believed to begin with changes in the genes in a single cell, a substance that can induce such a transformation in a bacterium's genes is presumed to cause cancer until further tests can prove otherwise. The test takes from a few hours to a few days to complete and costs between $300 and $1,000 for each chemical tested.[24] It is thus quick and relatively inexpensive and does not require the sacrifice of human or animal life. The Ames test, and other similar short-term procedures, allow rapid screening of many chemicals, including complex mixtures of air and water pollutants, and contaminants of blood and urine.

On the other hand, these tests also have their limits. In the early validation trials of the Ames test, about 90 percent of the known carcinogens tested induced mutations in bacteria; the remaining 10 percent did not.[25] Further, some chemicals that do not cause cancer *do* lead to changes in genetic material. In addition, short-term tests do not predict other kinds of genetic damage, and of course many dangerous substances have harmful effects other than cancer. Thus industry's argument that a negative short-term test means that a product is safe to market is unwarranted since other tests may reveal different dangers.

Animal tests provide more conclusive evidence but cost more and take longer. In this kind of study, mice, rats, dogs, or monkeys that are raised in a laboratory are exposed to measured doses of a substance suspected of being hazardous. At the end of a given time period, the exposed animals are compared to an unexposed group that has been kept in a similar environment. If the exposed group has a higher rate of illness, tumors, deaths, or offspring with birth defects, the substance is considered potentially dangerous.

Critics of animal studies level several charges against them:

- People are different from mice or rats, so we cannot generalize from one species to the other.
- Everything causes cancer; if we inject mice with vanilla ice cream, they will eventually get cancer.
- The doses given to test animals make it impossible to generalize to humans, who are never exposed to such high doses.

None of these criticisms is valid. Study after study has shown that a substance that causes cancer in one species is almost certain to cause it in another species.[26] The cancer may be different, however: a chemical that causes skin cancer in a mouse may induce liver cancer in a dog; it nevertheless leads to a malignant tumor in both. Furthermore, virtually every chemical known to cause cancer in humans (the only possible exceptions are benzene and arsenic) does so in experimental animals.[27] Thus, while animal studies cannot tell us either the *number* of cancers a toxic substance will induce or the target organ, they *can* predict whether a chemical is carcinogenic.

The media's constant discussion of cancer and the hazards in the environment has led many people to believe that anything can cause cancer—that we live in a sea of carcinogens. In fact, of the 7,000 substances toxicologists have tested on animals, only 1,500 (or 21 percent) have been shown to cause cancer.[28] As two science journalists observed, "The bottom line on carcinogenesis testing is this, You can drown an animal under a heap of it, or beat an animal to death with a sock full of it, but if it isn't carcinogenic you can't give an animal cancer with it."[29]

When a series of animal studies showed high rates of bladder cancer in rats who had been fed the artificial sweetener saccharin in amounts equivalent to human consumption of eight hundred cans of diet soda a day,[30] many people joked about the results. They missed the point: since most scientists agree that if a substance is carcinogenic, a small dose will lead to a low rate of cancer in an exposed population and a larger dose will lead to a higher rate, by giving animals high doses researchers can quickly establish whether there is any danger at all. Subsequent tests—at varying levels of exposure—can determine the effects of differing doses.

Animal tests have several advantages. First, using rats or rabbits to test toxic chemicals can be a way of preventing human exposure. For instance, animal tests on thalidomide, a drug that caused serious birth defects in European children, stopped it from being distributed here in the United States. Second, unlike the short-term Ames test, animal experiments make it possible to study health effects as diverse as cancer, liver or kidney damage, nervous system diseases, skin disorders, and reproductive problems. Damage to an animal organ is a clue to a substance's potential impact on humans. Finally, animal tests, though slower than short-term tests, are still relatively quick and inexpensive. Rodents are exposed for two or three years; humans are often exposed for twenty to thirty years before a cancer is detected. According to the National Cancer Institute, a carefully designed animal study for a carcinogen costs approximately $400,000.[31] In 1975 Dr. Samuel Epstein, an environmental physician, calculated that for about $210 million a year animal tests could be conducted on all the new chemicals put on the market each year. This would amount to only 0.2 percent of the chemical industry's yearly gross sales, or less than 4.9 percent of its annual after-tax profits. Looked at in another way, testing every new chemical would cost about 7 percent of what the United States now spends for medical care of cancer patients.[32]

The disadvantages of animal testing lie in the problem of generalizing from the laboratory to the world outside, a drawback that can underestimate the real danger to humans. Unlike experimental rodents or monkeys, people do not live in cages and eat regular diets. Their exposure to thousands of substances cannot be controlled. So while animal tests give us clues as to the kinds of damage a particular substance can cause and the relationship between the degree of exposure and its health consequences, they cannot tell us what will happen when a substance is released into the human environment. Nor can they tell us how toxic exposure, individual behavior, and past medical history will interact to influence health. Asbestos fibers and cigarette smoke, for example, work together to magnify the risk that each poses alone. Men who smoke are ten times as likely to die of lung cancer as nonsmokers and asbestos workers are five times as likely to die as other workers—but asbestos workers who smoke have a lung cancer death rate *fifty-three* times as high as other nonsmoking workers.[33] Unfortunately, only human studies can provide this grim assessment of risk.

Epidemiological studies compare patterns of health and disease in human populations. By looking at data from official health statistics, questionnaires and surveys, health examinations, and measurements of environmental pollutants, epidemiologists hope to find links between such agents as the tuberculosis bacillus, cigarette smoke, or high fat diets and such diseases as tuberculosis, lung cancer, or heart attacks. Their ultimate goal is to uncover clues that can help to prevent disease.

Animal and short-term tests are relatively easy to conduct, but the results cannot safely be generalized to humans. On the other hand, epidemiological studies can serve as a basis for public policy but they are enormously difficult to carry out. One problem is the long latency period for diseases like cancer. It is both time consuming and expensive to find the necessary data on exposure twenty or thirty years ago in order to relate it to a population's present health problems. It is equally difficult to keep track of a population for two or three decades into the future to see who gets sick.

A second problem is defining both the cause of the disease and the effect of the substance under study. Few toxic substances leave a unique imprint on the human body. To the pathologist, a lung cancer due to smoking looks no different from one induced by air pollution. Except in those rare cases where only one chemical is known to cause a condition (as polyvinyl chloride causes an unusual form of liver cancer and asbestos causes mesothelioma, a cancer of the lining of the chest cavity), clinical examination does not help to indict a specific chemical culprit. Effects are equally difficult to isolate since many environmentally related health problems can be caused by more than one agent. Liver damage, for instance, can result from exposure to carbon tetrachloride, infectious hepatitis, excess alcohol consumption, or any combination of the three. Unless epidemiologists can separate the exposures and document the effects, they cannot establish a causal relationship.

Statistical methods that can sort out these variables do exist, but they require immense amounts of accurate data, which are not routinely collected. Epidemiologists thus have a difficult time getting the information they need. While several states have excellent cancer registries, records of other health problems that may be caused by environmental factors do not exist. Only recently have state and local authorities begun to collect occupational and environmental exposure data. Further, since few doctors ask their patients about these exposures, clinical studies (reports on a series of patients with similar problems) rarely take into consideration the effect of toxic agents.

Another methodological problem that epidemiologists face is the difficulty of finding a control (or comparison) group that is similar to the population under study in all respects except its exposure to the agent in question. As we saw in the Love Canal case history, scientists criticized the EPA for comparing the health of Love Canal residents to that of others living in Niagara Falls, a city known for its industrial pollution. They reasoned that the EPA failed to find higher rates of illness at Love Canal because the entire city has a high incidence of pollution-related disease. This is an ongoing problem, since as more and more people are exposed to the myriad chemicals described in the previous chapter, it will become increasingly difficult to find a truly unexposed control group.

Another category of problem forces us to apply the epidemiologists' results cautiously. Epidemiology can only give information about the effects of past exposures. As Richard Doll and Richard Peto, two prominent British epidemiologists, note, "Unless epidemiologists have studied reasonably large, well-defined groups of people who have been heavily exposed to a particular substance for two or three decades without apparent effect, they can offer no guarantees that continued exposure to moderate levels will, in the long run, be without material risk." Since few studies meet these criteria, epidemiological research is rarely more than suggestive. Doll and Peto conclude that "the human evidence that is currently available does not allow us to express any confident opinion about the extent of the harm that the introduction of these [toxic chemical] substances may or may not do in the future."[34] Thus although epidemiological research can demonstrate that a substance is probably dangerous, especially when the risks are great and the population exposed is sizable, it cannot prove that an exposure is safe. Negative findings may mean that a substance is harmless but it may also mean that the methods used were inadequate to detect risk.

Despite their limitations, epidemiological studies are important because they alone document the toll of death and disease caused by toxic exposure. By looking at human activity in the complexity of the real world, epidemiologists can derive practical guidance to prevent future illness. But such research must not substitute for other tests because ultimately epidemiology depends on human experimentation: people who have been exposed to a substance are watched for ill effects. To industry officials, this seems to be a point in its favor. They frequently argue that regulation should be based on epidemiological data alone. For example, the Formaldehyde Institute, the trade association of formaldehyde manufacturers, has urged that because animal data on carcinogenesis are ambiguous, federal regulatory action should await conclusive evidence from human studies.[35] The medical director of the Exxon Corporation put forward an even more ghoulish argument for epidemiological research: "A regulatory program based on experimental screening models to evaluate new chemicals prior to their entry into the environment . . . will hinder the better documentation of this correlation [between human and animal carcinogenicity data]. When a carcinogen is prevented from entering the environment on the basis of screening results [from animal or bacterial tests], there can be no data regarding that exposure."[36]

Each of the methods of research—short-term, animal, and human studies—has its unique benefits and constraints. But all share a more fundamental problem rooted in the nature of modern scientific inquiry. In the last two centuries scientists have made tremendous advances in understanding the atom and its particles and the living cell and its chemistry. This progress has depended on investigators studying complex systems by breaking them down into their constituent parts, which they then scrutinize in isolation. While such reductionism, as it is called, has been helpful in building a base for modern technology, it limits the ability to provide answers to many of the questions that confront us. In the real world numerous variables interact within and among a variety of systems, so that the linear cause-and-effect model has left scientists understanding remarkably little of how biological systems work and of how humans interact with their environment. More complete knowledge of the impact of toxic chemicals on our ecology awaits major improvements in the sophistication of our research methods. In the meantime, we continue to depend on the inadequate methods we have.

The social and political context in which scientists work complicates the use of these methods even more. One problem is the inadequate resources—given the magnitude of our ignorance—available for environmental health research. Over 1,000 new chemicals are introduced every year; 20,000 to 50,000 dump sites scattered across the country contain unknown quantities of unknown chemicals; 279 pesticides are sprayed on food crops. If every scientist in the country were to devote his or her career to investigating environmental hazards to our health, there would still be many unanswered questions—and there are nowhere near enough trained environmental scientists to investigate even the major hazards.[37] And the federal government's failure to fund research adequately guarantees that the problem will continue. In fiscal year 1983, for example, the Department of Defense spent twenty-three times more than did the EPA on academic research.[38] Overall, the EPA got less than 1 percent of the federal government dollars allocated to campus research.

Another problem plaguing the study of environmental health hazards is the chaotic way scientific research is organized. Environmental scientists work for the government, for universities, and for industry. No national plan coordinates this research; resources are not allocated according to any rational or stated priority system. In the federal government alone, thirty-seven agencies have some responsibility for environmental research.[39] Many task forces and committees include experts from several agencies, but no one official has responsibility for coordinating the research.

University scientists are more interested in basic than applied research, and are usually funded by the government for this work. When academic researchers do turn to more practical questions, their work is often sponsored by industry, which imposes its own demands on the questions that are pursued. Only a handful of independent researchers study health problems posed by environmental hazards, and when they do—as Beverly Paigen discovered in her work at Love Canal—they can jeopardize future grants, promotions, and their professional careers.

Corporate secrecy and a handful of corrupt scientists further compromise the contribution science can make to health protection. For example, environmentalists battled in court for ten years to get access to industry studies on the toxicity of certain pesticides. Companies such as DuPont, Union Carbide, Upjohn, and Velsicol had refused to release their test data on the ground that it might reveal trade secrets to their competitors. Yet as an article in *Science* noted, "Even the industry's primary attorney admitted that concerns about such unfair competition are largely speculative."[40] In a Kafkaesque compromise settlement, in 1982, representatives of six environmental groups were allowed to review the data in a federal office building outside of Washington, D.C., on the condition that they take no notes and discuss what they saw with no one else.

In some cases industry has been shown to fake or conceal scientific research—and no one knows if these were the only cases, or simply the tip of the iceberg. In 1977 U.S. attorneys in Chicago gained an eleven-count indictment against Velsicol Chemical Company for concealing data on the carcinogenicity of the pesticides chlordane and heptachlor—but the company's attorneys won a dismissal of the case on procedural grounds.[41] The most spectacular case of alleged fraud, however, involved Industrial Bio-Test Laboratories (IBT), the nation's largest commercial toxicology laboratory. In May 1981, after a five-year investigation by the Department of Justice and the Food and Drug Administration, the president and three top officers of IBT were indicted in Chicago on eight counts of distributing falsified scientific research and then attempting to cover up the scheme.[42] IBT's test results had been used to obtain government approval for 212 pesticides, yet an EPA investigation found that virtually *all* of the results were scientifically invalid.[43] According to the testimony of a former IBT technician, employees were ordered to omit deaths of experimental animals from official reports. A malfunctioning water sprinkler designed to provide drinking water to the mice and rats being tested, and to clean their cages, frequently flooded them with feces and food, yet the higher mortality of these animals was ignored. The same technician testified that results of blood and urine tests for an arthritis drug were submitted, although they had never been performed.[44]

One defendant in the case, Dr. Paul Wright, joined IBT in 1971 after leaving a position as a toxicologist at the Monsanto Company, a chemical manufacturer.[45] At IBT he supervised the testing of one of Monsanto's antibacterial agents; after he found the product safe, Wright returned to Monsanto as manager of its toxicology program. As one Department of Justice investigator put it, "IBT became the largest testing lab in the country, because companies knew this was the place to get the results they wanted."[46] In October 1983 a federal jury in Chicago convicted three IBT officials of trying to defraud the government and some drug and pesticide manufacturers by covering up inaccurate research data.[47] However, many of the products "cleared" by IBT are still on the market, earning healthy profits for their makers. In that same month, the National Toxicology Program of the U.S. Department of Health and Human Services began an investigation of "sloppy and inept" research at the Gulf South Research Institute, another private laboratory that performs cancer tests for the government.[48]

An editorial in the prestigious British medical journal *Lancet* described an even more damaging way in which industry influences scientific research:

> Another alleged tack is for the firm, singly or in combination with like firms, to set up supposedly independent research institutes whose scientists always seem to find evidence to support the stance taken by the firm, despite massive evidence to the contrary. Thus, when some high-sounding institute states that a compound is harmless or a process free of risk, it is wise to know whence the institute or the scientists who work there obtain their financial support.[49]

This view of industry's role in environmental and occupational health research does not inspire confidence. And as corporations increase their funding of university-based researchers, environmentalists will need to scrutinize academic research more carefully. In recent years petrochemical companies have flooded the campuses with money.[50] DuPont has given $6 million to Harvard Medical School for a new genetics department. Monsanto has awarded a $4 million contract to Rockefeller University for research on photosynthesis. Massachusetts Institute of Technology will get more than $7 million from Exxon for combustion research. In each of these cases, patentable discoveries will revert to the company. A recent article in *Science* warned of the effect of the "academic industrial complex" on independent, unbiased research.[51]

Scientists employed by the government face another conflict of interest. In her study of the Love Canal disaster, Adeline Levine described how scientists employed by the New York State Department of Health consistently narrowed their definition of the scope of the problem and recommended less expensive solutions.[52] She attributed these decisions to pressure from their employer, the state government, which hoped to reduce the costs of remediation. When government scientists challenge their political superiors, they can get fired. For example, Dr. Melvin Reuben, former director of the Experimental Pathology Laboratory at the Frederick Cancer Research Facility, was forced to resign his post in 1981 after warning that malathion, the insecticide used to kill the Mediterranean fruit fly in California, was carcinogenic. His boss at the lab accused him of engaging in "controversies that have both scientific and economic impact."[53] In 1982 Dr. Irwin Billick, former director of the Division of Environmental Research at the Department of Housing and Urban Development (HUD), was dismissed because HUD found his work on lead poisoning "unnecessary." But according to the Environmental Defense Fund's chief toxics scientist, Dr. Ellen Silbergeld, Dr. Billick was fired "at a time when the [Reagan] administration is trying to discount the importance of lead in gasoline."[54] Dr. Billick was the only government researcher with detailed statistics on the correlation between lead in gasoline and lead in the blood of urban children. A Cornell University researcher whose EPA funding for research on the utility companies' role in air pollution was abruptly slashed in 1982 observed that "our group erred in making too much progress. The Office of Management and Budget wishes to leave the EPA and Congress dependent on industrial sources of information."[55]

Even when scientists are unswayed by external pressure, their conclusions often differ widely. For example, researchers disagree on how many people die from air-pollution-related cancer. After carefully reading the extensive epidemiological literature on the subject, Doll and Peto, the two British epidemiologists mentioned earlier, concluded that the combination of air, water, and food pollution caused about 2 percent (8,500) of cancer deaths in the United States each year.[56] But a review of the same evidence by Marvin Schneiderman, former chief statistician with the National Cancer Institute, and Nathan Karch, former senior scientist with the President's Council on Environmental Quality, attributed 21,000 lung cancer deaths each year to air pollution alone.[57] And as the Love Canal case history shows, scientists who studied health problems near the dump site disagree profoundly on the ill effects of that exposure.

Such widely divergent findings cannot safely be used as a foundation for policy or for regulation: using the highest estimate of risk might mean banning a substance that is in fact moderately safe and imposing economic hardship on manufacturers, workers, and consumers. Using the most optimistic assessment, on the other hand, might lead to unnecessary illness or death. Even then, the real issue is that what is "safe" for a scientist may not be so for a citizen. The scientific method is an inherently conservative process. By training, scientists are skeptical; they want conclusive evidence before reaching a decision. Relationships between variables can be proven only after numerous studies in a variety of settings. The body of knowledge that sustains science was built by slow, painstaking research, by the countless repetition of experiments, and by rejection of all but the most convincing evidence. Citizens, on the other hand, need to know whether the dump site down the road *might* hurt their children or whether their drinking water *might* cause cancer. They want a chemical to be proven safe beyond a reasonable doubt before letting it into their lives. A potential threat is sufficient ground for protective action. Waiting for conclusive evidence would mean death or disease for them or their children, a heavy price for scientific certainty.

Neither side is wrong. Good science does require convincing evidence that can withstand rigorous scrutiny. Scientific research is indeed a crucial component in the control of environmental health hazards—it is the only method for making systematic observations of the relationship between exposure and health. Further, scientists have accumulated a wealth of data on environmental health that citizens would be foolish to ignore, and they have developed methodologies that can provide answers to some of the life-and-death questions that people exposed to a toxic substance ask. Their research, as we will see in the next chapter, forms the foundation for what environmental regulation there is.

But public policy requires a different standard of judgment. Problems arise when government or industry (or scientists in their pay) tries to convert a political question into a technical-scientific one. Residents, too, sometimes expect the "experts" to solve their problems for them. But however valuable their contribution, scientists cannot make decisions about what to do about an environmental hazard. They can collect evidence to document a risk and they can assist in assessing existing information, but no scientific principle can help to decide how many cases of leukemia are worth the cost of a pollution control device or how many miscarriages balance the advantages of the aerial spraying of pesticides. These are moral and social judgments that must be dealt with in the political arena.

3. ENVIRONMENTAL REGULATIONS:
Who Is Protected?

Case History: Washington Heights, New York

On August 6, 1980, a tanker truck carrying propane gas developed a leak while crossing the George Washington Bridge from New Jersey into New York City.[1] For six hours fire and police officials tried to repair a defective valve. According to experts who visited the scene, a single spark might have ignited a fireball one-eight of a mile in diameter. Hundreds, possibly thousands, of people living in the densely populated Hispanic, Greek, and Jewish neighborhood might have been killed or injured. As it was, over 2,000 people were evacuated and New York City experienced the worst traffic jam in its history. Finally, two police officers who were former plumbers used a $4 hardware store valve to plug the leak.

Alarmed by this near disaster, the Washington Heights Health Action Project, a community group formed a year earlier to combat environmental health hazards in upper Manhattan, began to investigate the transportation of hazardous materials in their neighborhood. What they discovered only increased their fears. According to a 1979 survey by the New York–New Jersey Port Authority, the regional body controlling the George Washington Bridge, more than one thousand trucks carried dangerous materials across residential streets in upper Manhattan every day. Their cargoes included flammable substances such as propane, cancer-causing agents such as benzene, lung irritants such as ammonia, and hundreds of other toxic materials.

Who was responsible for overseeing the safety of these vehicles and their deadly loads? Everyone and no one was the answer Health Action Project researchers came up with. The federal Department of Transportation (DOT) is charged with enforcing the Hazardous Materials Transportation Act. Passed in 1975, this act authorizes the DOT to establish uniform regulations governing

60

the movement of hazardous materials in interstate commerce. Despite the new law, DOT statistics show that hazardous materials accidents leading to injury or death have increased steadily since its passage. Yet in 1979 the federal government employed only forty-seven inspectors to check the estimated 1.3 million motor vehicles that transport hazardous materials every year, and in spot checks in Mississippi, Pennsylvania, and Long Island more than 20 percent of the trucks failed basic safety inspections.[2]

The New York State Department of Motor Vehicles also has regulatory responsibilities, but only for those trucks registered in the state. These vehicles must be inspected annually, but if a truck owner has a fleet of more than twenty-five vehicles he can legally bypass the state inspection system and inspect his own fleet. The New York City Fire Department decides the city streets that trucks with hazardous cargoes may use. Firefighters also respond to accidents, spills, and explosions. Cutbacks in the fire department, however, have reduced the number of employees available for both emergencies and inspection duty. In 1980 the fire department had only one team with special training in combating hazardous materials fires. Although Washington Heights has the city's highest volume of trucks carrying toxic cargoes, the local firehouse had no special equipment for this kind of disaster.

The Port Authority of New York–New Jersey controls traffic on the George Washington Bridge, setting rules on the kinds of trucks allowed on its roadways. But since the Port Authority sees its primary mission as promoting industrial development in the region, it has developed a cosy relationship with the local trucking and chemical industries. In a meeting with a delegation of concerned Washington Heights residents, for instance, the Port Authority's hazardous materials expert extolled the honesty of truck owners: "They're as concerned about safety as we are," he said, "so we find we don't need to do much checking up on them."

The hazards posed by the transportation of radioactive waste have complicated an already chaotic regulatory scene even further. In 1977 the New York City Department of Health banned the shipment of high-level radioactive materials on city streets because such shipments endangered the health and safety of New Yorkers. At the instigation of several New York universities and of utility companies that owned nuclear power plants, the DOT issued regulations that superseded the city's ban, which they argued constituted an undue interference in interstate commerce. Advocates of the federal regulations suggested that a possible route for nuclear shipments in New York City was through Washington Heights and across the George Washington Bridge.

Faced with what they considered the double peril of toxic chemicals and radioactive waste, members of the Washington Heights Health Action Project organized a coalition of concerned citizens who had been evacuated on the day of the propane truck incident, delegates from local neighborhood and church associations, senior citizens, and antinuclear activists. The coalition circulated a petition demanding that the Port Authority ban from the bridge all trucks with hazardous cargoes not bound for New York City and establish a regular inspection system for those vehicles that must enter. At community meetings, coalition members explained the issue and the demands. The local school board, the community planning board, and the area hospital's advisory board all passed resolutions endorsing the coalition's demands.

Following a community forum in November 1980 at which scientists and elected officials spoke, two hundred people marched across the George Washington Bridge to present the petitions signed by two thousand residents to Port Authority executives. A support group of New Jersey environmental activists joined the demonstration. As a result, the Port Authority agreed to meet with a delegation from the coalition—an unprecedented act for an agency known for its lack of accountability.

A few months after the propane truck accident, the New York City Fire Department announced new regulations that would ban trucks with hazardous cargoes from streets in Washington Heights. While public pressure led to these new rules, it was not sufficient to persuade the city to enforce them. At a meeting with Washington Heights activists, a fire department official explained that since the city was certain its new rules would be challenged in court by the Port Authority or the trucking industry, it decided to first commission a study to examine whether they were justified.

Meanwhile, New York City's ban on radioactive waste was still in jeopardy. At a hearing on the proposed DOT regulations held at police headquarters, more than five hundred people crowded into the meeting room to voice their opposition. Testimony continued until after midnight. Nevertheless, in February 1981 the DOT issued the new regulations, overturning New York City's ordinance, along with those of other localities. New York City, several states, and a variety of environmental groups immediately filed suit to enjoin enforcement of the new rules.

For more than a year the coalition organized by the Health Action Project continued to meet, demonstrate, educate its neighbors, and pressure local politicians. Then in the summer of 1981, discouraged by their lack of progress in ridding their streets of the chemical peril and frustrated by their inability to force the city or the Port Authority to respond, some members of the group decided to change its focus. If they could not mobilize sufficient power to bring about changes in upper Manhattan, they could expand their definition of the problem and its solutions. They therefore organized a conference on hazardous materials in the entire metropolitan region, bringing together twenty labor, environmental, community, and antinuclear groups from New York and New Jersey. Among the issues discussed were toxic waste disposal, health and safety in the workplace, cutbacks in environmental protection, and "right-to-know" legislation. This regional approach permitted new and broader alliances and coalitions. For example, Washington Heights activists made contact with New York City firefighters worried about their exposure to chemical fumes, Newark residents fighting a Port Authority–sponsored toxic waste facility, and a Bronx group concerned about trucks carrying toxic cargoes through their neighborhood.

In upper Manhattan, however, hazardous materials were still crossing the bridge, and the local coalition had not been able to hold itself together. The cause that had sparked its efforts was not over. In August 1982 a U.S. Court of Appeals rejected the DOT's new regulations, thus upholding the New York City ban. In July 1982, however, the Long Island Lighting Company sent nineteen trucks loaded with enriched uranium through New York City without obtaining a certificate of emergency transport, the permit necessary for low-level radioactive shipments. The fuel was bound for the Shoreham nuclear power plant on Long Island. Six weeks later the New York City Department of Health charged the trucker with a violation that carried a maximum fine of $500— hardly more than a slap on the wrist.[3]

In September 1982 the New York City Fire Department began to *enforce its regulation banning trucks carrying hazardous chemicals from streets in Washington Heights. In a statement issued shortly before the gubernatorial primary election, New York City Mayor Ed Koch, a candidate for governor, stated: "I have promised the people of Washington Heights that they would not be imperiled by the transport of hazardous materials through the streets of their neighborhood. We are now giving notice that these materials will no longer be allowed on neighborhood streets."*[4]

More than two years after the propane gas truck incident, the coalition of concerned citizens had at least temporarily won its essential demand. In 1983, however, a federal appeals court reversed the lower court ruling and claimed that the DOT did indeed have the right to overturn the city's ban on radioactive waste transportation, and in early 1984 the U.S. Supreme Court upheld the DOT, thus clearing the way for radioactive waste to enter New York City.[5] *In addition, also in 1983, in response to pressure from community groups and elected officials, the New York City Police Department established a new unit to monitor and enforce hazardous materials transportation.*

Protection against toxic chemicals and other hazardous materials has depended primarily on a series of legislative acts passed in the last twenty years. The Clean Air Act of 1963, the National Environmental Policy Act of 1970, the Occupational Safety Act of 1970, the Environmental Pesticide Control Act of 1972, the Safe Drinking Water Act of 1974, the Toxic Substances Control Act of 1976, and the Resource Conservation and Recovery Act of 1976 are only a few of the bills signed into law during these years. In the first years of the 1970s legislators passed more and stronger environmental laws than any enacted in the past century. Despite this, the use of government regulations to control health hazards is plagued with difficulty. This chapter will examine how the government regulates environmental quality and how industry and community and environmental groups influence this process.

Public concern about environmental health was aroused in the early 1960s when a handful of scientists and organizations— Rachel Carson, Barry Commoner, the Union of Concerned Scientists, the Scientists' Institute for Public Information—began educating the public about the growing dangers posed by pesticides, radiation, and toxic chemicals. Increasing awareness spurred a popular movement for environmental protection.

From its beginnings this movement had diverse concerns: overpopulation, wildlife preservation, overproduction of consumer goods, and health-threatening pollution. Some organizations emphasized individual changes in life-style, while others worked for stronger environmental legislation and regulation.

Demonstrations, teach-ins, and other actions around the country on Earth Day, April 22, 1970, helped to convince politicians that a new movement—possibly as strong as the antiwar movement that had helped to force President Johnson out of office—was on the horizon. A leader of the League of Conservation Voters wrote in 1970, "Our goal is to prove that issues like ecology and population can decide the outcome of an election [so that] we can . . . succeed in defeating legislators whose policies are especially destructive to the environment."[6]

The panoply of new laws and agencies that emerged as a result of these efforts was both a victory for the fledgling movement and an attempt to contain and channel its energies. With the establishment of the Environmental Protection Agency (EPA), the Council on Environmental Quality, and the Occupational Safety and Health Administration (OSHA), environmentalism wrested an important foothold in the federal government, ensuring that the issue will remain on the national agenda. Moreover, the new agencies and regulations *have* helped to protect public health. The following are only a few of their many accomplishments:

- According to EPA studies, particulate air pollution fell by 33 percent between 1970 and 1975, while reductions in carbon monoxide emissions led to a 66 percent drop in the number of unhealthy days in big cities between 1976 and 1981.[7]
- A 1979 report by the Council on Environmental Quality estimated that mandated reductions in particulate and sulfur dioxide air pollution since 1970 were saving 13,900 lives a year.[8]
- In 1982 government researchers reported in the *New England Journal of Medicine* that between 1976 and 1980 average

blood lead levels in U.S. children dropped 37 percent, a decline they attributed to new rules ordering reductions in the lead content of gasoline during this period.[9]

- In 1983 the EPA released a study indicating that the level of PCBs in body tissues of Americans has declined dramatically since 1977, a drop attributed to the 1976 ban on the production of PCBs.[10]

As we saw in the previous chapter, scientific documentation of the health consequences following exposure to a toxic substance is imprecise, and respected scientists have contradictory interpretations of the data. Moreover, regulations are not issued by scientists (who can only document risks and assess the quality of the available evidence) but based on political judgments. To understand how these broader factors shape regulatory decisions, it is useful to examine how various parties influence the process of regulation.

Industry Influence on Regulation

Industry intervenes in a variety of direct and indirect ways to realize its goal of lowering the cost of complying with the regulations (so as to increase its profits) and of maintaining control over economic and social planning. Corporate leaders have developed imaginative techniques to make their case. For example, each year industry spends hundreds of millions of dollars on advertising against government regulations, almost always on the ground that such regulations will cost the *public* money. "Technology and large corporations are seen as important generators of growth, government as a critical impediment," proclaims an ad by Union Carbide, one of the largest chemical companies. "There is growing recognition," the ad goes on, "of the need to reduce the cost and uncertainty of government regulation."[11] In another advertisement, United Technologies, a major defense contractor, announces that "regulation imposes a growing burden on the economy and can exact costs that far exceed any benefits to society."[12] "Money spent to comply with regulation is money not available for investment to cut costs," observes Mobil Oil in a *New York Times* advertisement.[13] And the Monsanto chemical company, sends out a simple message designed to reassure the public:

"Without chemicals," they tell us, "life itself would be impossible."[14]

Recently, the chemical industry has become alarmed at the growing public mistrust of its willingness to protect the public's health. In his 1983 farewell speech as president of the Chemical Manufacturers' Association of America, William Simeral, an executive vice-president at the DuPont Company, said, "I submit that the fundamental cause of the chemical industry's public perception problem is basic fear of chemicals."[15] He complained that although his industry was not "poisoning America," the American public did not share that belief.

In response to these fears, manufacturers have placed advertisements stressing their commitment to safety and implicitly suggesting that government regulation is therefore unnecessary. "I'm a chemical industry scientist," the director of a major chemical company's toxicology department tells us in a newspaper advertisement. "When I test our products' safety I think of the people who'll use them. Including my family."[16] In mid-1983 Dow Chemical launched a $3 million public relations and research effort to "reassure those with concerns about dioxin," a contaminant of the herbicides Dow produces.[17] Dow is also a principal contributor to the Coalition for a Reasonable 2,4-D Policy, an organization with a $120,000-a-year budget that advocates continued use of 2,4-D, a herbicide used in Agent Orange and suspected of causing high rates of miscarriages in sprayed communities.[18]

The nuclear power industry has also used advertising to address its "public relations" problems. In 1983 one hundred companies, including nuclear suppliers such as the General Electric Company, engineering firms such as the Bechtel Corporation, and utilities such as Consolidated Edison, contributed more than $25 million to the Committee for Energy Awareness, an organization whose goals are "to create a political and public-opinion climate that will allow utilities to add new electrical generating plants when they are needed."[19] By its advertising the committee hopes to increase "the flow of positive news about electricity and nuclear power."

A second tactic is for industry to hire its own scientists and sponsor its own research. Thus if an independent investigator presents evidence that a certain substance is harmful, chances are that there will be a scientist on the manufacturer's payroll to testify that his or her work shows the opposite. Dow Chemical has spent millions of research dollars to prove its herbicides safe. At government hearings and in numerous court cases, Dow's staff scientists, or consultants hired for the occasion, have challenged the contention of independent researchers that phenoxy herbicides (2,4-D and 2,4,5-T) can cause spontaneous abortions and possibly birth defects.[20]

Sometimes an industry will even disavow other industry studies if they prove damaging. Research by the industry-sponsored Chemical Industry Institute of Toxicology found that formaldehyde increased the risk of nasal cancers in rats and mice. Such data are normally considered presumptive evidence for regulatory action, but in this case the Formaldehyde Institute criticized the study's methodology and argued that further evidence was necessary before action could be taken.[21]

Since environmental groups usually lack the resources to commission research, they cannot compete on equal terms with industry. At best, experts on opposing sides of the question neutralize each other. Then the supposedly "objective" regulators come up with a compromise between the two positions. Jacqueline Warren, a lawyer for the Environmental Defense Fund, described to a *Science* reporter how industry scientists influenced EPA regulations requiring manufacturers to maintain an inventory of hazardous chemicals:

> Throughout the deliberations about what the inventory should contain, EPA played the role of a neutral arbitrator of the disagreements between the chemical companies on one side and the environmental protection advocates on the other. The Agency appeared to have no views of its own or sense of its role in carrying out Congress's intent. This attitude has been apparent in other areas of EPA's jurisdiction in recent years, but never so blatantly as in the development of the inventory of hazardous chemicals regulations. As one who attended most of the public meetings, it was often the case on disputed issues that EPA seemed to be taking a head-count of the industry and environmental representatives to resolve the issue, and the tally was usually 50–3.[22]

Industry's success in forcing the EPA to see itself as a "neutral" judge of the various positions, rather than as an advocate for the people—its legal mandate—is one measure of its immense power.

EPA's respectful treatment of industry contrasts sharply with its lack of receptivity to the concerns of citizens. In an interview, an official in the agency's Boston office explained:

> We have a very strict policy that citizen groups can provide us with comments, discuss issues with us, but they cannot be our advisers. They cannot be decision makers along with us. We make the decisions. We make sure they understand where we're coming from and if they have comments to make, sure we look at them, and have all our bases covered. But they cannot expect to find the solution themselves. Government isn't going to allow that to happen.[23]

Corporate stables of research scientists hamper strict regulation in another way. By hiring environmental scientists at high salaries, corporations reduce the pool of talent available for government or independent research. For example, because of industry's allure to trained toxicologists, a specialty in short supply, the EPA had trouble finding toxicologists to monitor the Toxic Substances Control Act.[24]

Industry's influence on the regulatory agencies is not limited to the testimony of its researchers in congressional hearing rooms or courts. Lobbying on Capitol Hill, as well as within the EPA and other environmental agencies, affects both proposed regulations and those already on the books. Lobbyists for petrochemical and agribusiness interests meet regularly with influential legislators and appointed officials to discuss regulations that affect their industries. Generous campaign contributions from these same industries help to make these meetings mutually beneficial. For example, in 1982 twelve members of the House of Representatives who voted that year to weaken the Clean Air Act received $197,325 in campaign contributions from the political action committees of ninety-three corporations found in violation of the act's emission standards during that year.[25]

While many of the national environmental organizations have begun to hire lobbyists, big corporations and trade associations can always outspend and outnumber the public interest groups. Internal Revenue Service regulations that forbid tax-deductible contributions to lobbying organizations further hamper environmental groups' ability to raise money to influence legislation. As a result of such disparities, industry has easier access to those with power. For example, during the 1982 conflict between the Reagan administration and environmentalists over proposed changes in the rules regulating the amount of lead permitted in gasoline, Congressman Toby Moffett, chairman of the House Energy and Commerce Subcommittee on the Environment, released data showing that EPA officials held thirty-two meetings with representatives of the refining, lead, and petroleum blending industries, but no meetings with health or environmental groups before announcing the new rules.[26]

Industry lobbyists often find that the senior staff of the regulatory agencies listen to their plight with sympathy. Part of the reason for this friendly response may be that regulators often began their careers on the other side of the desk. Many high officials of the EPA, Nuclear Regulatory Commission (NRC) and other agencies concerned with the environment worked for years for the industry they are charged with regulating. For example, William Ruckleshaus, President Reagan's selection to replace Anne Burford after she was forced to resign as director of the EPA in 1983, had been a vice-president at Weyerhauser, a timber company with 1,300 pollution law violations between 1977 and 1982.[27] He also sat on the boards of at least four major companies that the EPA is charged with regulating.[28]

Nor is the flow of personnel from industry to regulatory agency a one-way street. Many agency officials have resigned to take high-paying jobs with the corporations they once monitored. Stewart Udall, secretary of the interior under Presidents Kennedy and Johnson, left office to set up an environmental consulting firm that helped electric utilities plan for the future.[29] John Rademacher left a high position at the EPA to become a vice-president of Velsicol Chemical Company.[30]

If public opinion advertising, sympathetic hired scientists, and lobbying fail to get industry its way, legal action is always another option. Corporations can and do challenge environmental regulations after they are issued. Once again, industry's ability to hire skilled lawyers and pursue expensive appeals makes it difficult for environmental groups to counter their impact on the government. Every occupational health standard promulgated by OSHA has been challenged in court by the affected industry, and during the years and years of legal wrangling industry continues to expose its workers to the suspected hazards.

When, on the other hand, government agencies or citizen or

environmental groups seek to use the courts to force an industry to clean up an environmental hazard, the corporation often uses legal proceedings to delay action for years. In February 1972 the U.S. Department of Justice filed a suit against Reserve Mining Company, a joint subsidiary of Republic and Armco Steel Corporations, for polluting Lake Superior with asbestos-contaminated taconite filings, a by-product in the manufacture of steel.[31] Public health officials had discovered high levels of asbestos particles in the drinking water of nearby Duluth, Minnesota, and feared that the taconite posed a major health menace to those who drank Lake Superior water. Yet it was not until April 1974 that the court ordered Reserve to find an alternative method of disposal, and it was not until 1980 that the company set up a land-based disposal system. Finally, in 1982 Reserve agreed to a settlement requiring it to pay $1.84 million to filter the drinking water of Duluth and three other Lake Superior towns.[32] Reserve's ability to hire lawyers and file numerous appeals thus delayed action for ten years. Dow Chemical used a similar legal strategy to delay a permanent ban on the use of 2,4,5-T.[33]

If these remedies fail, industry can turn to bribery, payoffs, corruption, or other illegal activities to get its way. Sometimes government officials encourage this tactic. George Guenther, OSHA director under President Nixon, suggested to the president's 1972 campaign staff that "four more years of properly managed OSHA" should be used "as a sales point for fundraising and general support by employers."[34] Many corporations use illegal dump sites as a cheap solution to their disposal problems. Among the regular clients of a waste disposal firm convicted several times of illegal dumping in Pennsylvania were General Electric, Exxon, Contrail, Bechtel Power, and DuPont.[35] In 1981 twenty-four companies, including IBM, General Motors, DuPont, and General Electric, agreed to pay $7.7 million to clean up 60,000 barrels of dangerous waste they had improperly dumped at a site in Seymour, Indiana.[36] In 1983 Boise Cascade agreed to pay a $100,000 fine to the Minnesota Pollution Control Agency because three of its employees had tampered with a state test of waste water from its paper mill.[37] And for every act of illegal dumping that is detected, many more go unreported. Small penalties and lax enforcement undoubtedly encourage corporate managers to risk fines rather than spend money on compliance with regulations.

Often illegal activity is difficult to uncover. Corporate executives and lobbyists meet regularly with government regulators, usually behind closed doors. When such meetings lead to pro-industry policies or enforcement practices, it is impossible to know whether a promised campaign contribution, an "honorarium" for a speech, or an offer of a future job helped to convince the official to change his or her mind. The revolving door between the corporate suite and the regulatory agencies ensures that both parties share a value system that protects such potentially illicit collaboration.[38]

The final tactic that industry may use to thwart regulation is to threaten to move abroad if a rule it opposes is passed. The threat is not an idle one. Asbestos production, for example, has moved from the United States to Brazil, Taiwan, and Mexico, countries with few environmental regulations. Hazardous waste companies are now building disposal sites throughout Africa and Latin America.[39] And despite the fact that lower wages and taxes and not strict regulations are the major motivation for overseas corporate journeys[40]—and that many industries have sizable investments and a skilled labor force here in the United States and so are less mobile than they pretend—in times of economic hardship few politicians, or communities, are willing to call industry's bluff in the face of environmental blackmail.

ENVIRONMENTAL BLACK MAIL

State and Local Regulations

As we saw in the Washington Heights case history, state and local governments also have regulatory responsibilities, often making it more, rather than less, difficult to protect the public's health. To begin with, these agencies encounter the same problems as do national regulators: scientific uncertainty and industry influence. And access to resources allows the corporations to put great pressure on local governments; here the threat to move to another town or state to escape strict regulation is real. Few local governments are able to stand up to such economic pressure. For example, General Electric successfully blocked every attempt to regulate its polychlorinated biphenyl (PCB) waste within New York State by threatening to close down its plant on the Hudson River. Not until national regulations on PCBs were passed did

GE's discharges into the Hudson end.[41] In another case, in 1983, Minnesota Mining and Manufacturing (3M) chief executive Lewis Lehr warned Minnesota Governor Rudy Perpich that passage of a state hazardous waste bill that would include compensation for victims of toxic chemicals would provide a powerful incentive for 3M to conduct its business elsewhere. When the governor signed the bill despite these threats, he was informed that the company's plan to expand its operations in the state were now "questionable."[42]

In addition, multiple layers of regulation can create such confusion that effective control is sacrificed. Small companies may have difficulty understanding which regulations apply to them or how to comply with several overlapping sets of rules. Residents who fear they are endangered are often referred from one agency to another. Joseph Cashman, a citizen of dioxin-contaminated Times Beach, Missouri, complained about the government's lack of coordination in assessing risk and taking corrective action. "Somebody has got to be in charge to get something done," he said. "Nothing is coordinated. They are all pulling in different directions [and] we cannot get any facts."[43]

The larger corporations can play one level of government off against the other. In response to growing fears of accidents involving the transportation of hazardous materials, many localities have banned the shipment of certain toxic materials through populated areas. Trucking industry associations have challenged these bans in court, arguing that the less stringent federal regulations should take precedence.[44] In some municipalities, as the case history showed, local regulatory agencies have not enforced existing rules for fear of industry lawsuits that are expensive to contest.

Resources for Regulation

Even when regulators have accurate scientific information, an explicit mandate from the legislatures, and cooperation from industry, their ability to enforce the regulations depends on having sufficient resources to carry out the job. Yet this is scarcely ever the case, as these examples show:

- In 1979 the Department of Transportation had only 222 inspectors for hazardous materials transportation by air, rail, water, and highway. Forty-seven inspectors were expected to check the 1.3 million motor vehicles that transport hazardous materials every year, and nine had to check the estimated 21,000 manufacturers of the containers used to transport toxics.[45]

- The Department of Agriculture is charged with monitoring pesticide levels in red meat and poultry. Yet the USDA labs where meat is tested have the ability to detect only 46 of the 143 pesticides and drugs known to be found in meat.[46] One USDA official told a reporter: "What we detect is only the tip of the iceberg—we are in the midst of jet-age problems but we are equipped with horse-and-buggy technology."[47] The Food and Drug Administration tests for pesticides in fruits and vegetables. Its labs can detect no more than 79 of the 279 pesticides registered for use on produce.[48]
- In 1979 OSHA had enough inspectors to visit every office shop, or factory covered by the law once every fifty years. Even by leaving many plants uninspected, OSHA could only check the most dangerous workplaces infrequently.[49] Since 1980 President Reagan's cuts in OSHA's budget have further reduced the number of inspectors.

Several factors determine the level of funding for environmental protection: the strength of the environmental movement, the power of industry, and the policies of the president in office. Corporate contributions to the campaigns of anti-environmental candidates are one way of ensuring that fewer funds are spent to protect the environment. Between January 1981 and February 1982, for example, five major corporations with poor environmental records (Dow Chemical, Weyerhauser, Republic Steel, Occidental Petroleum, and Amoco) contributed $144,800 to candidates for House and Senate seats. Their largest contributions went to four Republican senators, all of whom voted against a motion to restore funding cuts for the EPA in fiscal year 1982.[50]

Finally, many hazardous substances, including those generated by small businesses and those used as fuels in home and industry, are exempt from any regulation whatsoever. According to New Jersey's Representative James Florio, a member of the House Energy Committee, "Mounting evidence can only lead one to conclude that there is as much, if not more, waste that is escaping regulation than is currently covered under the law."[51]

Environmental Protection Under Ronald Reagan: An Endangered Species?

The environmental reawakening of the early 1970s by no means ended industry's influence on the regulatory process, but it did create a counterforce within the federal government. Many of the scientists and lawyers who worked for EPA, OSHA, and other environmental agencies were genuinely committed to enforcing the law. Political constraints may have limited what they could accomplish, but by collaborating with citizen and public interest groups they were able to win some important victories. The regulatory agencies became arenas in which industry and environmentalists struggled to make their case. Never before had environmental organizations had such an important forum.

The election of Ronald Reagan in November 1980 virtually ended even this limited federal role in environmental protection. By deregulation, defunding, and defederalization, Reagan changed the EPA, which had been a battleground for opposing political forces, into what one activist dubbed the "Industrial Protection Agency."

A review of the Reagan administration's environmental record illustrates its openly pro-business biases. Former EPA administrator Anne Gorsuch Burford proclaimed that "sound environmental regulation can only be as good as the scientific foundation upon which it is based."[52] Yet the EPA's 1983 research budget was only 60 percent of what it had been when Burford took over the agency. Research programs on acid rain, health monitoring, and toxic chemicals were cut or eliminated because they could have produced data contradicting official positions. Since Reagan's election, the EPA has proposed relaxing environmental standards for carbon monoxide (a cause of serious lung and heart diseases), lead (a cause of brain damage and learning problems in children), hazardous waste, and several other substances. Intense public opposition has thwarted some of these efforts, but as an EPA staff member observed, "These people [officials] aren't going to regulate until they see dead bodies."[53] In January 1983, for instance, a federal judge finally ordered the EPA to publish air pollution standards for arsenic, a carcinogen, within 180 days: the agency should have issued the standard two years earlier, a delay that the judge said "flouted the considered judgment of Congress."[54]

The EPA has also drastically reduced its enforcement actions.

In 1981 the agency referred only sixty cases of violations of all its pollution laws to the Justice Department for prosecution. The previous average—itself inadequate—had been two hundred cases a year.[55] In 1981 the EPA filed only two cases for enforcement of hazardous waste regulations; fifty-five cases had been initiated in the last eighteen months of the Carter administration. As Jeff Miller, a former EPA enforcement chief, observed, "The civil servants perceive the new team doesn't want enforcement and they're afraid for their jobs, so the safest thing is to do nothing."[56]

Reagan has also fought for major revisions of the Clean Air and Clean Water acts. His proposals would delay compliance dates, exempt large sectors of industry from certain regulations, and weaken protection of public health. In 1983 the new EPA administrator, William Ruckleshaus, announced his support for weakening these two major regulatory programs.[57]

Reagan's top appointments at the EPA guarantee that the corporate foxes will have chicken for dinner every night. EPA's former chief, Anne Burford, previously a Colorado legislator and telephone company lawyer, had once sued the EPA to exempt Colorado from certain requirements of the Clear Air Act.[58] John Daniel, EPA chief of staff, served as a lawyer for Johns-Manville, a major producer of asbestos. He had also worked for the American Paper Institute, where he coordinated an industry effort to weaken the Clean Water Act in 1972.[59] The people Reagan appointed to replace the top EPA management he fired in early 1983 were no less pro-business. For example, during his tenure at Weyerhauser Company, the world's largest forest products concern, Ruckleshaus had given a speech in which he called on the EPA to "abandon or at least modify its traditional role as an advocate for a cleaner environment and instead adopt the role of educator."[60] In a 1981 letter to Vice-President George Bush, Ruckleshaus stated: "As the first administrator of EPA, I am largely to blame for most of the bad things done to business by environmental regulation over the last decade. Given my culpability, I am mightily interested in seeing constructive change in our country's social regulatory policy."[61]

The harmful effects of Reagan's transformation of the EPA are most immediately apparent in the agency's failure to correct pollution hazards. In May 1981, for example, EPA officials told regional offices to stop ordering cleanups of allegedly hazardous waste sites without prior approval from Washington.[62] The main office then rejected half these requests, forcing scores of communities to remain exposed to toxic chemicals. In one case EPA officials acknowledged that for two years the agency had failed to pursue information that could have forced such petrochemical companies as the Allied Corporation, the Olin Corporation, and Rollins Environmental Services (a waste disposal firm) to clean up a toxic lagoon in southern New Jersey. In 1983 an administrator in the New York regional office admitted that "in retrospect, it was a poor decision" not to further investigate the site, listed by EPA itself as one of the most dangerous in the nation.[63] In another case, congressional investigators charged the agency with delaying the cleanup of the Stringfellow Acid Pits near Riverside, California, in order to harm the senatorial campaign of Governor Edmund Brown.[64] Coincidentally, one of the companies accused of illegal disposal of chemicals at the Stringfellow site was Rohr Industries, where Reagan adviser Edwin Meese III had been vice-president.[65]

The Reagan administration also gave industry an unprecedented role in making local enforcement decisions. Rita Lavelle, assistant administrator for waste programs (later convicted of perjury), sought to negotiate settlements with polluters rather than force them to clean up under the law.[66] And in 1983 EPA's Chicago regional administrator testified before a congressional subcommittee that his superiors in Washington had ordered him to allow Dow Chemical to edit a report on its role in polluting the soil in Midland, Michigan, with dioxin.[67] Dow deleted the report's conclusion that it represented "the major source, if not the only source" of dioxin in nearby waters.

Cost-Benefit Analysis: The New Religion

President Reagan's policies have damaged the environment and weakened the protection of our health. While it will take decades to measure the full impact of Reagan's record, hundreds of thousands of people have already been exposed to dangerous levels of toxic chemicals as a result of his efforts. But the philosophy of regulation underlying the Reagan administration's action, although more explicit in its pro-business tilt, does not differ significantly from that of his predecessors. Shortly after taking office, for instance, Reagan issued an executive order proclaiming that "regulatory action shall not be undertaken unless the potential benefits to society for the regulation outweigh the potential costs to society."[68] In 1978 President Carter had made a similar proclamation; earlier still, President Ford had ordered that all environmental regulations had to have an "inflationary impact statement."[69] Finally, the 1980 Supreme Court ruling that OSHA must make a quantitative estimate of risk levels as a prerequisite for the regulation of the carcinogen benzene provided a judicial mandate for cost-benefit analysis in environmental regulation.[70]

The call for cost-benefit analysis has thus been heard in three administrations. Its support by a broad spectrum of corporate and political leaders makes it a cornerstone of capitalist philosophy. Yet cost-benefit analysis requires its users to accept a number of assumptions: that all the costs and benefits of a proposed regulation can be ascertained with a high degree of certainty; that all the costs and benefits can be assigned a dollar value; that all interested parties can agree on the costs, the benefits, and their dollar values; and that all the costs and benefits accrue equally to the entire population.

Each of these is subject to challenge. First, as the review of methods of scientific research in chapter 2 indicated, our knowledge of the health effects of toxic substances is still very limited and subject to different interpretations: even when competent researchers review the same set of data, their estimation of the benefits of regulation can differ by a factor of one hundred or one thousand.[71] Thus it is difficult to know what the real benefits will be. Second, the costs of complying with a regulation are equally difficult to predict, although for different reasons. When the federal government proposed a new occupational standard for vinyl chloride in the early 1970s, the plastics industry charged that the standard was beyond its ability to comply with and that its enforcement would lead to the loss of 2 million jobs. Nevertheless, OSHA stood firm and no jobs were lost, while the cost of compliance was only 5 percent of the $65 billion the industry had estimated.[72] Similarly, in 1975, during congressional debate on the Toxic Substances Control Act, an industry-hired consultant argued that the cost of complying with the act would be prohibitive, far higher than the government estimate. Yet when members of Congress asked to see the cost data that formed the basis for the industry estimate, they were told that the figures had been destroyed by prior agreement among the companies.[73]

In order for costs and benefits to be compared, they must be reduced to a common measure, usually dollars. For example, a proposed pollution control device might cost $100 million to purchase and install in every factory producing a toxic substance, while the failure to install the equipment is predicted to cause one hundred additional deaths from cancer. The benefit of the device is then calculated by adding the costs of hospital care, funeral expenses, the lost wages of the victims, and so on. This grisly mathematical exercise poses several problems. Since unknown costs or benefits cannot be assigned a dollar value, the calculations use incomplete data and generally underestimate benefits. Furthermore, analysts disagree on what value to assign a given cost or benefit. How much is human life worth? Estimates range from $158 million, used in one occupational health study, to $10,000 a year, the amount society is willing to pay to preserve the life of an unknown person in a public nursing home.[74]

And how can present costs be compared with future benefits? Leslie Boden, a Harvard University economist, has observed that it is easier to estimate short-run costs than long-run ones.[75] In addition, while it may well cost less in the long run to design new plants and equipment to reduce air pollution, few corporations are willing to trade high present costs for uncertain future benefits. Moreover, many events cannot be assigned a dollar value. What is the real cost to a family and to society of a child who dies of chemically induced leukemia? What payment can compensate the individual whose chromosomes have been damaged by exposure to toxic chemicals?

Most important, costs and benefits are not equitably distributed throughout society. A corporation's costs for a pollution control device may yield a benefit for those living near a plant: they will get sick less, often live longer, and pay less for cleaning materials dirtied by pollution. If these benefits deprive the company's shareholders of dividends worth more than the total value of the residents' savings, does this mean the control devices should not be purchased? The formula for comparing costs and benefits does not take into account *who* benefits and *who* pays. And how to weigh the corporate costs against the public health benefits is a political decision.

The limitations of cost-benefit analysis have led many to question its usefulness as a tool for making regulatory decisions. Three scientists who worked with the EPA to develop the standards for carcinogens in drinking water put their criticisms this way:

> The agency has not found cost-benefit analysis to be a powerful tool, at least for developing a regulatory strategy to deal with the presence of organic carcinogens in drinking water. The agency is inserting cost-benefit analysis into the adversary process of rule-making. It is predictable that both (or all) sides will be able to introduce analyses that support their points of view. Each will then bring forward analysts, of equivalent credentials, who will argue among themselves about the superiority of their particular analyses in a language all their own. Incomprehensible analysis will then drive out understandable argument. The cost-benefit analyses are likely to capture only a small part of the problem and deflect attention from the rest. This will not contribute to better decision making.[76]

A review of cost-benefit analysis by the House of Representatives Subcommittee on Oversight and Investigation of the Committee on Interstate and Foreign Commerce concluded that the "limitations of the use of benefit/cost analysis in the context of health, safety and environmental regulatory decision-making are so severe that they militate against its use altogether."[77] Boden warned that cost-benefit analysis is "an evaluative method that may bias political decisions against even those regulatory decisions that are cost-effective."[78]

If cost-benefit analysis is subject to such serious criticism, why has it been endorsed by three presidents, the Supreme Court, and most corporate leaders? Why has it become the new religion of regulators charged with protecting the environment, workers, and consumers?

David Noble, an MIT professor and historian of technology, suggests one answer. "While the demand for cost-benefit analysis and economic cost accounting appear reasonable enough, in that it is typically voiced as a simple plea for rationality," he wrote in the *HealthPAC Bulletin* in 1980, "it is in reality more than this; it is a political strategy designed and fostered, quite deliberately, to undermine, stall and hamstring the regulatory agencies."[79] Or as Burke Zimmerman, former staff scientist for the House Subcommittee on Health and the Environment, put it, the call for cost-benefit analysis is the "invention of those who do not wish to

regulate, or to be regulated. Its primary use in government decision-making is to avoid taking action which is necessary or desirable in order to truly protect the health of the public or the
integrity of the environment."[80]

In response to growing criticisms of cost-benefit analysis, some
of its proponents have tried to give it a more democratic veneer. In
1983, for example, EPA Administrator William Ruckelshaus
asked the residents of Tacoma, Washington, to participate in the
agency's deliberations on regulating carcinogenic arsenic emissions from a copper smelter. At a public forum Tacoma's citizens
were asked whether they wanted to close the smelter, depriving
the community of 575 jobs, or risk several additional cancer
deaths each year.[81] Ruckelshaus explained, "I am asking for comment from the people most directly affected by the risk from exposure to arsenic, after first informing them of the nature of that
risk."[82] He went on to note that the "right to be heard is not the
same as the right to be heeded. The final decision is mine." Yet
such attempts at public involvement fail to address the basic problems of cost-benefit analysis. As Ruth Weiner, an environmental
chemist from Western Washington University who has studied
the Tacoma smelter, observed, "It is up to the EPA to protect
public health, not to ask the public what it is willing to sacrifice
not to die from cancer."[83]

In short, the demand for cost-benefit analysis is an attempt by
corporations and their political supporters to define the terrain on
which regulatory decisions are made. The complicated mathe-
matical formulas and the arcane terminology exclude ordinary
people from the decision-making process. The reliance on scientists and economists makes it appear that regulatory decisions are
technical, not political. The use of dollars as the common measure
ensures that economic considerations will take precedence over
concern for such immeasurable values as health and social justice.

Cost-benefit analysis thus obscures the basic question of who is
to decide the risks that we are to face. It makes popular debate and
political action more difficult, and meanwhile its proponents
maintain decision-making power for themselves. As long as cost-
benefit analysis remains the cornerstone of regulatory philosophy,
the U.S. government cannot be trusted to protect public health
and the environment.

4. ARMING YOURSELF FOR BATTLE:
How to Get Information on Environmental Health Hazards

Case History: Staten Island, New York

Curious about the new construction going on near his home in a wooded area of Staten Island in 1972, John Quinn asked some of the workers what they were building.[1] "A home for aged Norwegian seamen," was their reply. But the two large circular excavations hardly looked like a future rest home and Quinn soon learned that they had lied to him. What was being built were two 900,000-gallon storage tanks for LNG, or liquified natural gas.

Determined to find out more about LNG, Quinn and two of his neighbors began to look for information. They found out that LNG is a liquid form of the natural gas used to heat their homes and cook their food. When stored, it is concentrated 600 times and cooled to $-260°F$; under these conditions, a spill can create an explosive volatile cloud. In his research Quinn came across an article on LNG in a maritime publication. The article, written by a coast guard lieutenant commander, described the potential effect of an LNG supertanker accident on the Arthur Kill, a waterway separating Staten Island from New Jersey. Such an accident, under certain weather conditions, could create a gas cloud over Staten Island; if ignited, the gas could burn back to its source, incinerating everyone and everything in its path.

It was not the first time that Staten Islanders had faced an environmental problem. The smallest and least developed of New York City's five boroughs, Staten Island had long been considered a refuge from city life. In 1843 Henry David Thoreau visited the island and admired its "very fine scenery." But the very qualities that attracted nature lovers and those fleeing Manhattan—unspoiled land and cheap property—also made Staten Island desirable to industry. By 1960 industrial development had made farming on most of Staten Island impossible. Once the source of a large strawberry crop, now even the flowers were pocked with holes corroded by sulfur dioxide air pollution, the by-product of a copper reclamation plant across the Arthur Kill.

Staten Island was also New York City's garbage dump. By the mid-1970s the island's Fresh Kills marsh had become the largest landfill on earth—each day the Department of Santitation dumped more than 10,000 tons of solid wastes there. Other city dumps on the island had illegally accepted toxic chemical wastes; when this scandal was uncovered in 1982 Staten Islanders demanded city action to clean up the hazard.[2]

So when the LNG issue arose, it did so in a community that already had some experience in combating environmental problems. The men who had done the preliminary work turned to their parish priest for help. When he heard their story, he urged his congregation to begin a campaign to protect their community. He also asked two parishioners with a history of civic involvement to look into the problem in more detail. These two people, Edwina and Eugene Cosgriff, began by attending a meeting sponsored by Distrigas Corporation, the company that owned the tanks. At that meeting residents got Distrigas officials to place a few drops of LNG into a puddle of water so as to confirm their claims as to its safety. When the puddle spontaneously ignited into a fireball, the Cosgriffs decided they had proof enough of the dangers of LNG and that it was time to take action. Their first task was to learn as much as they could about LNG, the tanks, and the agencies responsible for their regulation. They began to write letters—to city, state, and national elected officials, and to every regulatory agency they could think of. The response was buck-passing. "Regulation of gas tanks is a state responsibility," claimed the Federal Power Commission. "We know nothing about it," said the state agencies, "perhaps the city can help you."

A few of the letters included thick government reports. The Cosgriffs and those they had recruited spent hours poring over these documents, taking careful notes. They learned that an LNG explosion in Cleveland, Ohio, in 1944 had killed 133 people, injured hundreds of others, and devastated fifty square blocks. They also discovered that the New York City Fire Department had no emergency evacuation plans in the case of such a disaster.

At the suggestion of their priest, the Cosgriffs asked a number of concerned citizens and neighborhood groups to form an organization to fight the tanks. They christened the new group BLAST—

Bring Legal Action to Stop Tanks. BLAST members began to educate their neighbors about the LNG tanks, sending speakers to meetings at churches and synagogues, Parent-Teacher Associations, civic organizations, and political clubs. They also distributed pamphlets and wrote letters to the local newspapers.

In early 1973, frustrated by their failure to stop construction, Edwina Cosgriff wrote a "woman to woman" letter to Pat Nixon, then the First Lady, asking for her help. Mrs. Nixon forwarded the letter to the New York regional director of the Environmental Protection Agency, who wrote a reassuring reply, saying that the LNG tanks posed no more risk "than the gas tank in your car."

On the very day Edwina Cosgriff received the letter, disaster struck. A smaller LNG tank, built some years before on a different part of the island, exploded and forty workers were killed. Edwina Cosgriff went to the site of the accident to read the EPA letter to reporters.

The tragedy turned BLAST's trickle of support into a groundswell. Within a few months the group had collected 35,000 signatures on a petition demanding that the tanks under construction not be filled and empowering BLAST to represent the signers in negotiations with government or industry. Three hundred and fifty women demonstrated with infants in strollers in front of the tanks, stopping cement trucks from making deliveries. Other demonstrations targeted Staten Island Borough Hall and City Hall. At one point 150 residents took a five-hour bus trip to Washington to attend a Federal Power Commission hearing on the gas tanks.

BLAST began to hold weekly meetings, a practice that was to continue for five years, an impressive record for any environmental group. Sometimes only twenty-five people showed up, but whenever a show of strength was needed more than five hundred would attend. Mothers of young children, fearing for their families' safety, were the organization's backbone. Staten Island college students, recently turned on by Earth Day and the environmental movement, helped with research and picketed. Their professors provided technical assistance and expert advice. Older residents joined because they were worried by the threat to their peaceful neighborhood.

Staten Island is home to one of the oldest black communities in New York City, settled shortly after the Civil War. Its residents were oyster fishermen until pollution ruined the oyster beds. They too opposed this new danger to their community. A local union of correctional officers, whose members were mostly black, supported BLAST because the detention center where they worked was located in the shadow of the tanks. Construction unions, however, favored the gas tanks because they hoped for new jobs.

The mass media also helped to educate the public, especially after the 1973 explosion. A Daily News reporter wrote a glowing account of BLAST's work after Gene and Edwina Cosgriff showed her government reports and scientific articles that demonstrated that the gas company had misrepresented LNG's safety. A local television station did a four-part special on LNG that brought a flood of telephone calls and letters to the BLAST office. Several journalists wrote books on LNG. One, a fictionalized account of an LNG explosion in New York City, was sent to every member of the New York City Council. A more scientific volume provided further ammunition for BLAST's pamphlets and flyers.

As the situation on Staten Island began to receive national attention, scientists and other experts began to offer support. From the beginning, two local college professors had helped to analyze government reports and had spoken at public meetings and agency hearings. In 1974 an MIT professor and cryogenics expert, James Fay, testified at a Federal Power Commission (FPC) hearing that LNG facilities should not be allowed in populated areas. A former official of the gas industry appeared at a 1975 FPC hearing to refute his colleagues' claims about LNG's safety. In 1977 the Congressional Office of Technology Assessment issued a report that concluded that "it is generally agreed that if the vapor from a large LNG spill ignited, it would be beyond the capabilities of existing fire fighting methods to extinguish."[3] A year later the General Accounting Office, after sending its experts to consult with BLAST members, also released a report condemning the storage of LNG in populated areas.

As BLAST mobilized more and more people, first on Staten Island, then in nearby Brooklyn and Northern New Jersey, politicians began to flock to their meetings. In public, all local elected officials supported the group's position. Turning promises made in front of television cameras into action proved more difficult, however. On Staten Island, U.S. Congressman John Murphy vowed that the tanks would never be filled—but in Washington, a

Staten Island high school class that had heard a talk by Edwina Cosgriff happened to sit in on a subcommitee hearing at which John Murphy spoke in support of continued federal subsidies for the LNG industry. Both the oil and shipping industries contributed heavily to Murphy's campaign.

Other politicians were more consistent. For example, in 1976, after strong lobbying by BLAST the New York State Legislature passed a law banning LNG tanks from densely populated areas. Staten Islanders fought a hard but successful battle to exclude a clause that would have allowed the existing tanks to remain in operation. Later the New York City Council enacted a regulation requiring that existing tanks be recertified every two years. Since the Staten Island tanks could not meet these new city standards, they were not eligible for recertification.

BLAST has now existed for more than a decade. What has it accomplished in this time? First, not a drop of liquified natural gas has ever been stored in the two tanks that were built in 1972. BLAST and its supporters have prevented the use of a facility that was already under construction and they succeeded in passing on the cost of mothballing the tanks to the owners' stockholders rather than to the ratepayers.

Second, BLAST has helped to pass state and city laws that not only banned filling the Staten Island tanks but also protected other communities in the state that might face similar threats in the future.

Third, BLAST has provided technical and political support to other New York City groups fighting LNG tanks and to concerned citizens in other states and even as far away as Canada, England, and Scotland. MASS BLAST, a New England offspring of the Staten Island group, won restrictions on the transportation of LNG in the Boston area.

Fourth, the Cosgriffs and their neighbors have built an organization that has existed without interruption since 1972. BLAST has maintained a presence that has so far been able to protect its victories. The group changed from a protest organization into a sophisticated political force with influence at the local, state, and national levels.

Finally, BLAST has led to lasting changes in the way many Staten Islanders think about their environment, their political system, and their own power to make changes. At the start of the struggle, activists thought that the safety arguments would convince everyone, but this hope was soon crushed. Utility companies and politicians seemed more impressed with political power than with logical arguments. "We didn't want to get involved in politics and economics," said Edwina Cosgriff, "but we were forced to."

This involvement changed the participants' views of politics. "As a result of our activities," Cosgriff told an interviewer, "I don't have much respect for politicians. Politicians are much more concerned with people who contribute to their campaigns than with voters." BLAST members learned how to translate their new understanding into results. "Once you've made it a political issue," observed Cosgriff, "you can't sit back. You have to keep people on the politicians' backs."

This distrust spread to industry as well. "We asked the company, 'Why did you build here on Staten Island?'" said one member of BLAST. "All they told us was, 'Because it is cheaper.' Big business is money, and they don't care about people."

For some Staten Islanders these new insights led to cynicism, apathy, or depression. But for others the long struggle brought optimism. Edwina Cosgriff believes that BLAST's campaign "gave a lot of hope to people who thought you couldn't fight City Hall." Now, she said, "People tell me, 'I feel I can get involved and make a difference.'"

The two empty LNG tanks still stand on the edge of a swamp on Staten Island, less than two miles from the Cosgriff's home. Deep cracks in the empty tanks have led BLAST to initiate new legal efforts to force their removal. Recently, the gas industry made a proposal to use the tanks as a "peak sharing facility" to store extra fuel during periods of high demand. BLAST opposes this, fearing a precedent that could allow the tanks' use for other fuels. Its members remain cautiously optimistic: they believe that their decade of experience has given them the strength and skill they need to continue to protect their community.

Across the country, community groups have learned about the environmental hazards that face their neighborhoods in a variety of ways. For some the hazard was obvious. "You can't miss it," noted a member of a group fighting toxic waste disposal in Tennessee. "The river is dirty, the air smells, and people complain about health problems." In Pittsburgh, previous disaster had alerted the public to the dangers of air pollution: in 1948 a period of intense air pollution in Donora, Pennsylvania, killed 20 people and sent nearly 6,000 more to the hospital.

For others the discovery has been less direct. An activist in Newark, New Jersey, first found out that a toxic waste facility was planned in his neighborhood when he read an article about it in a technical journal. Similarly, soon after the Three Mile Island accident an environmental group in Massachusetts got an anonymous phone tip about a secret nuclear reactor on a nearby army base. They investigated and found that although the reactor was not operational, other military activities on the base were releasing radiation into the environment. In New Mexico, a group of citizens became alarmed about air pollution after seeing aerial photos that showed plumes of smoke emanating from the industrial plants in their area. In the state of Washington, residents began to notice dead fish in a previously unpolluted stream and then discovered that the fish kills followed local commercial spraying of herbicides.

However a hazard is discovered, the first stage of a campaign against it is to gather information. This includes answering questions such as: How much and what do you need to know about an environmental hazard in order to fight it? What are the best sources of information? What obstacles can you expect in the search for information and how can these obstacles be overcome?

What Activists Need to Know

Community groups fighting environmental health hazards need information to determine whether there is a danger, to assess the claims of those who deny that there is one, to educate the community about the hazard, and to mobilize people and plan a successful campaign to end the danger.

Each of these tasks requires a different kind of information. To demonstrate that the community does indeed face a health hazard, activists should ideally be able to identify a hazardous substance, the routes by which that substance comes into contact with humans, and the evidence that the substance has in the past damaged health. Possible sources of information include personal observation, environmental testing, previous research studies, interviews with scientists, and government records. In the early stages of the investigation, suggestive (rather than conclusive) evidence may be sufficient. Like the detective checking into a suspicious death, activists are at first trying to decide whether to begin a more systematic inquiry. At this point clues that will stand up in a court of law are not necessary; leads that will provide a more complete picture of the problem are what is needed.

Communities that suspect an environmental health hazard are usually faced with one of three situations. In the first, a potential hazard (a dump site, a nuclear power plant, high-voltage power lines) is present and residents want to know if it has already caused health problems. What they need is evidence that the hazard has been associated with health problems in the past. In the second situation, people observe what seems to them an unusual disease pattern (a high rate of miscarriage, several cancer cases, many skin rashes). They need to ascertain that this pattern is indeed unusual, to determine what is known about its possible environmental causes, and then to search for its source. In the third situation, there is both an obvious environmental hazard *and* an unusual disease pattern. The task is to collect data that link the two. In practice, then, the kind of information needed to document the danger will depend on which of the three situations pertains.

Corporations or government agencies responsible for environmental health hazards often try to deny that there is a problem, and since they have greater resources than most community groups they can more easily gather and present to the public the information to make their case. To assess these claims environmental activists need to look at the adequacy of the industry's argument and to search the scientific literature for contradictory evidence. Good sources for this kind of information are sympathetic scientists, national environmental organizations, popular scientific books, technical journals, and friendly industry or government officials. Disproving just *one* of an industry's major contentions can strike a powerful blow at its credibility. For example, the 1981 discovery that the Pacific Gas and Electric Company (PGE) had made serious design errors in bracing California's Diablo Canyon nuclear power plant against earthquakes helped the antinuclear Abalone Alliance discredit PGE's claim that the plant was totally safe.[4]

Still another type of information helps to educate people in the community. Here what is necessary is a clear, simple description of how the threat affects people's day-to-day lives—information that will help people to protect themselves, both as individuals and as a community. On the other hand, to mobilize people for action requires evidence that will make them angry, concerned, or fearful, as well as information that will make them believe that they can act to eliminate or reduce the danger. Sources are health professionals, books, and other environmental groups.

Finally, organizers need to have the right kind of information to plan a successful strategy to get rid of the hazard. They need to know how serious the problem is; whether there is an alternative solution, and if so what kind; the political vulnerabilities of those responsible for the hazard; their potential allies and enemies; and the tactics that have worked in similar situations. Other environmental and community groups can often help to answer these questions.

The goal of the research is not to find out *everything* that is known about a particular hazard—that could be a lifetime job. Nor is research a distinct stage with a beginning and an end: activists cannot wait to act until they have "complete" information because at every stage of an environmental struggle there will be new questions and new problems to research. The goal is to gather the evidence needed to make the decision about what to do next.

In some cases preliminary research may convince concerned residents that their suspicions are unfounded. If the investigators find evidence that truly reassures them, they need to say so to the public. It is irresponsible to alarm people unnecessarily; false charges can only damage the credibility of the environmental movement. With so many real hazards, time and energy should not be wasted on spurious ones.

In any case—and this is the key point—information is rarely the factor that decides who wins or loses an environmental struggle. True, BLAST's successes at government hearings and in the courts depended in part on its accurate information and careful research. But its ability to prevent the LNG tanks from being filled was essentially a *political* victory. Data and facts are only one part, although an important one, of the outcome of scientific and regulatory controversies. The political power of the contestants, the prevailing social climate, and the institutions in which the battle takes place all play a critical role in the final decision.

Further, activists who believe that those who have the most information will win are doomed to failure: big corporations and government agencies will *always* be able to buy more facts. The environmentalists' strength is not in the thickness of their files or the academic credentials of their supporters; it is in the quality of their evidence and its relevance to people's concerns, and in their ability to use it to protect people's health and improve their lives.

Sources of Information

Environmental organizations can find information on a hazard from a variety of sources. This section describes some of these and gives examples of how activists have used such evidence to plan their campaigns. Other sources can be found in the resources section at the end of this book.

Citizens

Perhaps the most common starting point for community action against an environmental hazard is one person's observation of a change in his or her surroundings. Paying attention to the environment and listening to the neighbors are key parts of being an environmental activist. For example, residents in a town in northern California first learned that their properties were being sprayed with pesticides when they awoke to the sound of helicopters. Foul odors from a factory in Connecticut encouraged its neighbors to investigate whether the bad smells were also toxic. Bad-tasting water in Hardeman County, Tennessee, led to the discovery of an abandoned toxic dump site. In each of these cases citizens followed their ears, eyes, noses, or tongues to find a serious hazard. By listening to peoples' complaints and by asking them if they are having problems, activists can gather the raw material for pursuing an investigation of a suspected hazard.

Systematic records of these perceptions of unusual health problems or changes in pollution levels can then be used to generate hypotheses that can be tested in more formal health surveys. Not every opinion as to what caused a health problem will be correct, but by putting together many peoples' experiences it is sometimes possible to discover a relationship between exposure to a pollutant and health.

Residents with special skills can sometimes help to discover a problem. An amateur mineralogist in New Jersey realized he had come upon an abandoned radioactive dump near his home when his Geiger counter suddenly went crazy. A physicist in Massachusetts wanted to try out a new spectrometer—a device used to analyze the organic chemical content of a substance—and tested a sample of drinking water from his home. He was shocked to find it contaminated by carcinogens such as dioxane and trichloroethylene.[5] His discovery led him to initiate a battle to clean up his town's water supply.

Once a suspected hazard has been identified, ordinary people can be involved in documenting the extent of the problem. In Anderson County, Tennessee, children riding their bicycles were asked to keep an eye out for illegal dumpers of toxic waste. Whenever they saw a truck headed for the landfill they told their parents, who in turn notified a local citizens' group.

The New Jersey Public Interest Research Group developed a more systematic way of involving citizens in gathering information.[6] Groups of residents conducted "stream walks" by foot or canoe. These expeditions would look for pipes discharging waste into a waterway, collect samples for laboratory analysis, and record fish kills or other signs of pollution. Their findings were gathered into a report documenting industrial water pollution in southern New Jersey.

Involving local residents in gathering information serves several purposes. First, it expands the number of people monitoring health and the environment beyond a handful of activists and government regulators. The more eyes, ears, and noses testing for pollution, the more likely a hazard will be discovered early. Second, community expertise—the knowledge people have of their own surroundings—can often supplement professional expertise. The Love Canal residents' knowledge of local underground streams, for example, helped scientists to test the hypothesis that disease patterns were related to toxic chemicals in these streams. Finally, organizers use community involvement to become acquainted with the people they seek to mobilize. Working together to gather information can set the stage for an ongoing relationship of mutual trust.

Government Agencies and Government Records

A group in the Piedmont region of North Carolina concerned about toxic waste in their area found out about local dump sites by reading a congressional report on the problem. Research in the files of the U.S. Securities and Exchange Commission (SEC) helped a group in Newark, New Jersey, learn more about previous environmental violations of a company planning to build a toxic waste facility in their neighborhood. Reports of county environmental officials in California gave an environmental organization there data on which nearby oil refineries were discharging wastes into waterways. These three examples show how activists can use government agencies and records to gather evidence on potential health hazards. But learning what information is available and how to get it can be a time-consuming procedure. Table 1 lists various governmental sources, describes the kind of information they can provide, and suggests how activists can use it. However, since each locality and state has its own agencies and procedures, community researchers will need to discover which of these sources are relevant in their area.

Environmental groups use government reports for several purposes. One is to identify polluters in order to target them for action. The permits that corporations must file with regulatory agencies can often assist in this task. Regional Environmental Protection Agency (EPA) offices, for example, maintain files of all National Pollution Discharge Elimination System (NPDES) permits. These list the substances every factory is permitted to discharge into a waterway. Since EPA enters the information into a computer, it is possible, for instance, to get a list of every industry discharging into a specified river. By comparing the toxicity and volume of the pollutants, the major polluters can be identified and action taken against them. Using this method in combination with the stream walks described earlier, the Clean Water Action Project of the New Jersey Public Interest Research Group found several corporations in violation of their NPDES permits to discharge into the Passaic River. At the same time they found several leaking dump sites near the river that did not meet state licensing require-

TABLE 1   Sources of Government Information for Environmental Activists

	Agencies	Kind of information	Possible uses
Local	Health	Vital statistics (births, deaths by cause, reportable diseases)	To compare cancer death rates or infant mortality in two communities; to look at changes in death or disease rates over time
	Environmental protection	Records of air and water pollution levels at monitoring sites; citations for violations of local pollution laws	To monitor pollution levels in various neighborhoods or changes over time; to check for past violations of suspected polluters
	County recorder, assessor, or clerk	Records of land and building ownership; records of actions of governing body	To learn who owns a toxic dump site or polluting factory; to check government contracts; to look for evidence of improper influence; to find out local laws
	Regional air pollution, pesticide, or hazardous waste boards	Records of licenses for emissions, spraying, or dumping; history of complaints against a factory or industry; violations of pollution control laws	To investigate a suspected polluter's record of violations; to learn prevailing practices in the area as a basis of comparison; to check if a permit is being violated
State	Health and/or environmental protection	State implementation plan for compliance with Clean Air Act; Environmental Impact Reports; violations of state health or environmental codes	To investigate the record of a suspected polluter; to determine environmental impact of a proposed project; to explore opportunities for legal challenges to a proposed plan
	Agriculture	Pesticide spraying permits	To investigate the relationship between spraying practices and illness; to find the names of substances being sprayed

State Government Departments (continued)	Forestry	Herbicide spraying permits	As above
	Transportation	Violations of hazardous materials transportation regulations	To document lack of enforcement; to target trucking companies with multiple violations
	Economic development	Permits for siting of industrial projects	To explore the opportunities for legal challenges
	Energy	Siting of conventional and nuclear power plants	To determine if a utility company has met state and federal standards for siting and construction
	Labor	Toxic substances used in various industries (under right-to-know laws)	To learn the harmful effects of exposure to industrial pollutants or wastes
Federal Agencies	Environmental Protection Agency	Permits for discharging into waterways; toxic chemical waste disposal permits; records on pesticide safety; lists of abandoned chemical dump sites; technical reports on many environmental hazards	To target polluters or illegal dumpers; to locate dump sites; to survey existing literature on an environmental hazard
	Occupational Safety & Health information	Records of violations of the Occupational Safety and Health act; standards for occupational exposures to toxic substances	To investigate a company's safety and health record; to identify issues for labor-environmental coalitions
	Nuclear Regulatory Agency	Records on safety violations in nuclear power plants	To determine whether accidents, spills, etc., have occurred at a local nuclear power plant
	Food & Drug Administration	Records of violations of Food & Drug Act	To identify pesticides found in food samples
	Department of Transportation	Violations of laws on hazardous materials and nuclear waste transportation	To document violations of law in specific geographic areas

TABLE 1   Sources of Government Information for Environmental Activists (continued)

	Agencies	Kind of information	Possible uses
Federal Agencies (continued)	Securities & Exchange Commission	Annual reports of corporate activities, including lists of environmental laws and lawsuits	To examine the past record of a company suspected of pollution
	Centers for Disease Control	Reports of epidemiological investigations of environmental health hazards	To demonstrate the link between exposures and health effects in other communities
	Congressional Office of Technology Assessment	Investigations of technological issues facing Congress that often include assessments of the effectiveness of government programs	To summarize the literature on suspected hazardous technology
	Congressional General Accounting Office	Evaluations of federal programs	To document problems in federal programs

ments. These too were reported to enforcement agencies. The EPA has published a roster of 56,000 firms that lists their location and the types and quantity of toxic waste they handle. With this information from Resource Conservation and Recovery Act permits, groups can identify local generators and disposers of toxic waste products.

A second use for official data and reports is for challenging industry or government contentions of safety. The Nuclear Regulatory Commission (NRC) is charged with keeping records on safety problems at nuclear power plants. Antinuclear groups frequently use NRC data to buttress their case against the nuclear power industry. In 1981, for example, Critical Mass, a Washington-based safe energy group, used power company reports filed with the NRC as a source for a nationally publicized study of the growing rate of accidents in these plants.[7] As was mentioned earlier, reports by the Congress's General Accounting Office and its Office of Technology Assessment helped BLAST to educate Staten Islanders about the dangers of LNG. BLAST's flyers frequently included quotes from these authoritative sources, thus enhancing their credibility.

Local agencies can contribute information. In Scottsdale, Arizona, concerned residents checked the complaint files at the regional Pesticide Control Board. When they found nearly nine hundred complaints, mostly from their neighborhood, they were in a strong position to challenge the spraying company's contention that there was no problem.[8]

Government reports can also help link a suspected hazard with a health problem or, more generally, provide scientific background material. In a campaign against lead poisoning in children, for example, the Washington Heights Health Action Project used data from two New York City agencies.[9] The Department of Health provided a list of the addresses of lead-poisoned children; by marking these addresses on a map, it was possible to select three blocks with clusters of poisoned children for education and organizing. Then the group used the files of the Department of Housing Preservation and Development to find housing code violations in the apartments where these children lived. Giving the tenants concrete evidence that negligent landlords caused both bad housing and sick children helped organizers to build a coalition of housing and health activists. In another example, Coastal

Carolina Crossroads, a group trying to block construction of an oil refinery in North Carolina, used a study by the National Cancer Institute (NCI) to support its claim that those living near refineries had higher rates of brain and other cancers. The NCI report was summarized in flyers distributed in the community, where it added weight to the arguments of those opposing the plant.

A few practical suggestions may help researchers use government records. First, it is often useful to look for what is *not* in the report. Many states require that the public agencies that fund, implement, or approve a new project file an environmental impact report (EIR) before taking action. EIRs must be made available to the public for comment and criticism. Many community groups have asked independent scientists or engineers to review an EIR for gaps or weaknesses in the government's case. For example, a lawsuit by a coalition of New York environmental groups that has so far successfully blocked a major highway planned on Manhattan's West Side depended primarily on the group's claim that the state's EIR ignored the risks to striped bass living in the Hudson River. In other situations, the failure to find on file a government permit or license has helped a group to track down an illegal polluter.

Second, researchers need the imagination to put data collected for one purpose to another use. Securities and Exchange Commission (SEC) files, for example, traditionally help investors and regulators make financial decisions. But the Greater Newark Bay Coalition Against Toxic Waste combed SEC files to check the environmental record of the SCA Corporation, a giant waste disposal firm that wanted to build a new toxic waste incinerator in Newark. SEC forms must include an annual report of any legal action against a corporation. When coalition researchers found that SCA had been investigated for toxic dumping in Edison, New Jersey, had a plant shut down by a judge in Illinois, was indicted for price fixing in Georgia and bribery in Ohio, and had been cited by the New York State Department of Environmental Conservation for illegal dumping, it was not difficult to convince their community that SCA's proposals for Newark required careful scrutiny.[10]

Third, environmental groups need to know where to look for government reports. Most federal agencies have regional offices with libraries and files. The EPA, for example, has ten offices around the country where citizens can find its major reports. Such

Washington-based agencies as the SEC and the NRC will respond to written requests for information. If they refuse, citizens have a way to pursue the information, which will be described later in this chapter.

The *Federal Register* prints proposed and newly implemented government regulations and occasionally includes a review of the research that justified the regulatory action. It is available in most large public and university libraries, as is the *Congressional Record*, which includes transcripts of all House and Senate hearings. In addition, every agency usually prints a list of its publications and the Government Printing Office distributes a catalog of all of the federal government's publications.

Organizations experienced in working with an agency can be helpful with finding state or local records. Groups like the Audubon Society or the Sierra Club usually know the ropes at state departments of environmental protection, while community groups can often assist in getting through the maze at city hall. Sympathetic government officials can also help track down information. For example, a group in Fort Worth, Texas, that was fighting a company polluting the air with sulfur dioxide found that an official in the air pollution control division of the health department was willing to give it information on the inadequacy of his department's enforcement procedures and monitoring equipment.

Occasionally, every effort to get information from the government will be stymied. In fact, resistance to requests for information is a problem activists encounter frequently. "Often the regional public health department deliberately obstructed our information search and made it quite a lengthy and worrisome process to obtain information," reported a member of a group fighting illegal toxic waste disposal in Tennessee. An American Friends Service Committee project based in Georgia said that "stonewalling by certain government agencies" blocked their access to accurate data on the transportation of radioactive materials. Sometimes government agencies simply refuse to give out information; at other times they give out what concerned residents believe to be false information. "The United States Forest Service repeatedly told us that 2,4-D and 2,4,5-T were harmless. They lied," noted a group in Northern California.

Several tactics are useful in forcing bureaucrats to release information. Elected officials can sometimes help. A Kentucky group opposing the continued storage of nerve gas on an army base asked the local Congressman to get information from the army. Congressional committees, state legislatures, and local councils will sometimes subpoena government records. Congressional investigations of the EPA in 1983, for example, revealed its failure to act to clean up local dump sites.

In 1966 Congress passed the Freedom of Information Act (FOIA), which requires public access to most of the records held by federal agencies. Many community groups have successfully used the act to get useful evidence. One antipesticide group in the state of Washington filed 256 FOIA requests to get EPA and Forest Service records on pesticide use. Friends of the Earth used the FOIA to get information from the NRC that showed that the Three Mile Island nuclear plant had dumped radioactive water into the Susquehanna River.[11]

A group can initiate an FOIA request simply by sending a letter describing the information needed to the appropriate agency. Community groups can ask for a waiver of the usual fees for processing the request. Should the agency refuse, its decision can be appealed. Unfortunately, several important kinds of information are exempt from the requirements of the act. Agencies are not required to release information on national defense, internal agency matters, records that would reveal industry "trade secrets," or investigatory files.

Recently President Reagan has proposed further restricting the public's access to government records. The administration argued that processing FOIA requests was too expensive, endangered national security, and threatened confidential corporate information.[12] In 1981 Attorney General William French Smith promised that the Justice Department would defend federal agencies against court challenges for information they wanted to withhold.[13] As a result of this attempt to undercut the Freedom of Information Act, environmental groups have joined journalists, civil liberties organizations, peace groups, and others to protect the public's access to government documents. So far the coalition has prevented changes in the legislation, but executive agencies are contesting increasing numbers of requests.[14]

In 1983 the Department of Energy proposed new rules to restrict the amount of public information on a broad range of nuclear power and nuclear weapons issues. The new rules were denounced by several governors, major trade unions, environmental organizations, universities, and libraries because they believed the rules violated the free interchange of information and compromised their ability to learn about and correct health and safety problems.[15]

Several states also have laws requiring the disclosure of public records that environmentalists have used successfully to gain access to the information they need. For example, state right-to-know laws can be used to obtain information on the names, properties, and toxicity of suspected hazards. A concerned citizens group in Pennsylvania reported that that state's right-to-know law "opened a lot of doors" to evidence that helped it to block a proposed landfill.

Scientific and Medical Experts

Scientists and medical researchers are probably the most useful allies an environmental group can have in its search for information. Since their professional training prepares them to find out what is already known about a subject, to generate new hypotheses and then to test them, they are ideally suited to help community organizations investigate environmental health hazards. But, as we shall see, consulting scientists can also cause problems. This section describes several levels of interaction between concerned residents and scientists, and then discusses some of the issues that arose in these encounters.

A good place to start learning what scientists think about environmental problems is to read some of the books they have written for the general public. Rachel Carson's classic work on pesticides, *Silent Spring*,[16] Barry Commoner's books on the environmental crisis—*The Closing Circle*, for example[17]—and Samuel Epstein's *The Politics of Cancer*[18] and *Hazardous Waste in America*[19] all provide activists with both specific information and a framework for understanding the broader causes of the problems facing their communities. These books also help in finding answers to questions that might be asked during an educational campaign.

Once community researchers have mastered the general issues, they can move on to more technical sources. The scientific literature on environmental health is growing rapidly, and includes journals and magazines specializing in environmental sciences (*Archives of Environmental Health*), public health (*American Journal of Public Health*), engineering (*Chemical and Engineering News*), and medicine (*New England Journal of Medicine*); textbooks; conference proceedings; and reports of professional societies. Journals are indexed and librarians in university or large public libraries can teach how to use them. Major libraries can also do a computer search for all the recent publications on a given topic. For a fee of $15 to $20, they will provide a computer printout of the title, author, and main conclusions of all relevant recent articles, which can then be found in the library. Many community groups ask college students to help them with this kind of research.

In this stage of the investigation the more precisely researchers know what they are looking for, the easier it is to find the answers. To learn everything already known about pesticides and human health, for example, would require years of work, but to learn what is known about a specific substance being sprayed in a community—say aldicarb or malathion—and its effect on reproduction or kidney function is far simpler.

Although a good research plan helps to focus library activities, promising but unsuspected leads should not be ignored. When Bonnie Hill, a schoolteacher concerned about the high rate of miscarriage in the Oregon town of Alsea, read a scientific article on the effects of dioxin on monkey reproduction, she came up with the hypothesis that the pregnancy losses might be related to the spraying of 2,4,5-T, a dioxin-contaminated herbicide.[20] Eventually her investigation forced the EPA to ban 2,4,5-T. Like a good detective, a researcher will pursue each clue until cause and effect have been linked as firmly as possible.

Another way to become up to date on current scientific knowledge is to attend professional meetings or conferences. For example, an antipesticide group in Washington sent a few of its members to a symposium on pesticides at which several nationally known experts summarized their work. The lectures and workshops saved the group weeks of time in the library. In another case, members of a northern California group working to curb toxic dumping by the local petrochemical industry attended a conference of cancer researchers. A paper reporting a relationship between high cancer death rates and the concentration of industry in that county provided the group with data that it used to educate trade unions and community residents.

Some groups communicate directly with scientists. For instance, the group that wanted to force an army base in Kentucky to get rid of nerve gas stored there wrote or telephoned nonmilitary specialists in chemical warfare to ask them to comment on the army's claims of safety. The Greater Newark Bay Coalition Against Toxic Waste met regularly with doctors from an occupational medicine clinic to discuss the safety levels and health effects of the toxic chemicals that were to be stored in a proposed waste facility. Later these physicians agreed to testify against the facility at hearings sponsored by the New Jersey Department of Environmental Protection. Establishing personal relationships with those with expertise in the relevant disciplines can help a group get the information they need to counter industry or government arguments. It can also increase a group's credibility in the mass media, the legislature, and the courts.

Finally, scientists can provide community groups with such direct services as testing for pollutants, conducting health surveys, and analyzing government reports. A few environmental groups pay scientific consultants, but most find sympathetic experts who are willing to volunteer their time. Technical assistance need not come only from nationally recognized scientists. High school chemistry teachers and their classes can test water for pollutants; biology classes can assess the impact of chemical contamination on aquatic life. What scientists can contribute is a methodology for the systematic collection of data that can then provide more convincing evidence than the anecdotal impressions of community residents.

One such tool is the health survey, in which residents are questioned about health problems. When the results of the interviews are tallied, it is possible to calculate the rate at which a certain condition occurs in the population under study. If this rate is significantly higher for a population exposed to some environmental toxin than for some unexposed comparison group, then it may be concluded that the exposure is causing a health problem. A health survey conducted by the Ayers City Fair Share chapter in Lowell, Massachusetts, convinced local officials to meet with the group to discuss the cleanup of an abandoned toxic waste site.[21] A second, more rigorous survey helped the group describe a pattern of disease in the neighborhood around the dump and showed that people living there could regularly smell chemicals, demonstrating an airborne exposure. The survey also helped to convince local newspapers that a problem existed: the Lowell *Sun*, for example, did a five-part series on the dump that included interviews with survey respondents who had health problems and had agreed to talk to a reporter.[22] Finally, as a result of the survey, the department of public health agreed to conduct a more extensive study. As we saw in the discussion of epidemiology in chapter 2, however, factors such as sample size, other possible exposures, and the demographic characteristics of the population may either exaggerate or mask a causal relationship. For this reason, health surveys by community groups are not a substitute for formal epidemiological studies.

The Love Canal Homeowners' Association also surveyed the health of neighborhood residents. Lois Gibbs explained that the results of the survey "allowed us to make sensible decisions about our families, about postponing pregnancies, whether we should move our children into relatives' homes, what questions we should ask our physicians and about what really happened in the neighborhood. We used the survey as a tool for ourselves, to receive media attention, to educate the public, and to move the health agencies to look into our community for further problems."[23]

Whatever the purpose of a health survey, community groups need to consult scientists for help in designing a valid study. Those who can provide assistance include epidemiologists and statisticians from universities or medical centers, social scientists experienced in survey research, physicians, and sympathetic public health officials. Sometimes a group's sense of urgency prevents it from taking the time to find professional help. This is almost always a mistake because a poorly done survey will be discredited by industry and the time spent in data collection will have been wasted. After conducting health surveys that provided no results, some groups have concluded that their efforts were a misuse of their time. If a group does not have the resources or technical assistance to carry out a careful survey, it should probably choose other activities to realize its goals.

Despite the contributions that scientists can make, involving them in the search for information can cause problems. Neither they nor concerned residents always know the limits of science's domain. In their quest for precision, scientific experts can confuse rather than clarify, and some concerned residents are so intimidated by technical information that they hand over all responsibility to the "experts." Scientists can exacerbate this problem by using jargon or by establishing a separate relationship with other scientists involved in the conflict. Scientists sometimes dominate the public debate on environmental hazards, usurping the role of the public. This abdication of responsibility not only ignores the political dimensions of the problem, but it leaves the community unable to monitor the situation when the scientists leave.

To avoid such problems, groups should decide exactly what they want from scientists before consulting them. If they establish a "contract" at the start, each party's role is clear. Some organizations have eliminated the problem of technical jargon by passing out whistles at the start of each meeting. Every time a scientist (or group member) uses a term someone does not understand a whistle blows. Everyone learns to speak clear English very quickly.

Even when helpful scientists can be found, there may be a frustrating lack of relevant scientific knowledge. A Massachusetts activist involved in a fight to force the army to stop burning uranium put it this way:

> It is hard to find people who know about this particular substance and who know how to figure out whether it actually poses a real threat to the health of our community. There are lots of "scientists" who have general theoretic knowledge about hazardous substances. [But it is] almost impossible to find people who have a down-to-earth practical knowledge that could help us figure out whether the particular practice of burning U-238 is actually a health hazard.

Lack of good research reflects the political context in which scientific priorities are set. It will be a long-term struggle to convince scientists, and the government and private agencies that fund their work, to create a body of knowledge that concerned citizens can use. Nevertheless a problem on which *no* information exists is extremely rare. By using the full range of sources described in this chapter most activists will be able to find some evidence to guide their work.

Sometimes a group will discover that a scientist involved in a dispute is biased. For example, those fighting herbicide spraying in Alsea, Oregon, joined with county and state officials to establish an independent scientific commision to investigate the suspicion that the spraying was causing miscarriages. Soon after the scientists began their work, the group learned that one of the physicians on the team had recently testified at an Oregon state hearing that the herbicide 2,4,5-T was safe.[24] Before that the man had been a paid consultant to the timber industry. The group exposed his record and looked for a replacement.

It is critical, however, to distinguish between scientists who are biased or present misleading information and those who honestly believe that the necessary information is not known. In the first case, the group simply needs to find a more independent scientist. In the second, the group needs to find other kinds of evidence to make its case. It doesn't make sense to behead the scientist-messenger who brings the bad tidings that a question cannot be answered.

Industry

Although industry often has an adversarial relationship with community environmental groups, many activists have garnered useful information from their opponents. Some of the techniques have been imaginative. A group fighting toxic dumping in Tennessee learned about the specific practices of one such facility by taking a tour of the plant with a high school class. An activist from another Tennessee group overheard the foreman at a toxic chemical disposal site telling his drinking companions in a local bar about an illegal spill. Newspaper reports of an industry petition for a waiver of an air pollution standard warned a Minnesota group of the threat of dirtier air.

Industry publications can help environmentalists in a number of ways. Exposing the contradictions between a company's public relations releases and reality damages its credibility. Trade or manufacturers' associations sometimes issue handbooks that are useful guides to standard practices. An operator's manual for pesticide applications and a safety book for truckers of hazardous materials provided groups fighting those threats with good background information. Package labels on pesticides have helped groups to learn about the dangers of these substances. Industry symposia and conferences are also useful. Opponents of a dump site in Pennsylvania learned about the technology of toxic waste disposal by attending an industry-sponsored forum on solid and hazardous waste.

Perhaps the most useful source of industry information comes from officials who sympathize with some of the goals of environmental activists. Discussions with "closet environmentalists in industry" helped a group in Dallas, Texas, discover the source of the lead contaminating its neighborhood. The Staten Island group, BLAST, received invaluable technical assistance from an executive of a gas company. Much of the technical information that has made the antinuclear movement such a formidable foe of the nuclear power industry has come from a handful of renegade officials who resigned their jobs as they came to understand the dangers of nuclear power.

If a group is fortunate enough to find someone who is or has been in industry and supports the group's goals, it must be careful to protect its ally. If an official plans to keep working for the industry, the group must find a way to disguise its source. Sometimes the corporation can be forced to acknowledge publicly what the group already knows privately; in other cases other sources can be used to confirm what the sympathetic official has suggested.

Often industry is reluctant to release information to environmental groups. When a group in the state of Washington requested information on the safety of a particular pesticide, the corporation replied that such data were trade secrets. The Long Island Lighting Company has consistently refused to provide opponents of the proposed Shoreham nuclear power plant with the information they have requested.[25]

Activists have several options for pursuing evidence that an industry denies them. First, they should be sure such information is not available from a public source—for instance, the files of the Securities and Exchange Commission, citations by the Environmental Protection Agency, or financial data included in the company's annual report. Then they can sometimes turn to former employees. The manager of a Tennessee landfill agreed to testify on the toxic chemicals dumped there because he was outraged at the lies in the public relations releases put out by the company managing the site and at the fact that he had not been warned of the danger to his own health.

Direct action, or the threat of it, will sometimes convince a reluctant company that they have more to lose by withholding information than by releasing it. Since most corporations are concerned about their public image, they will seek to avoid media coverage of a demonstration accusing them of covering up a health hazard. On the other hand, if there is a serious hazard and they know it, corporate executives may be willing to risk bad publicity rather than expose a costly problem.

The News Media

The news media can help environmental activists in two ways. They can publicize issues and struggles, which educates the community and increases the pressure for corrective action (see chapter 6). They may also provide information that can assist organizers in identifying a hazard and planning action against it.

A mother who led the fight to rid her child's school in Harlem, New York City, of flaking asbestos first learned of such a hazard from a radio report on a similar problem in a New Jersey school. Conversations with news reporters gave information on past illegal activities of the dumpers to Long Island environmentalists trying to force the cleanup of a town dump that had been contaminated with industrial toxic waste. In addition, most newspapers maintain files of past stories. A few days in the clipping morgue can save time in charting the history of a problem. It can also acquaint a group with earlier organizing efforts, leading to contacts with potential allies.

Not surprisingly, reporters are sometimes more successful than concerned citizens at getting government or industry reports. At Love Canal, for example, journalists consistently learned of the results of blood and other tests before citizens did. If a community group has a good relationship with a reporter, the two can trade information.

Finally, journalists, like scientists, sometimes write popular books on environmental hazards. Former *Niagara Gazette* reporter Michael Brown's book on hazardous wastes, *Laying Waste*,[26] and science reporter Paul Brodeur's books on asbestos and microwaves[27] provide facts and arguments useful for community education.

Labor Groups

Workers, first and hardest hit by toxic chemicals and other hazards, have often collected evidence on environmental threats to health. Even if such data document occupational, rather than community, exposures, activists can still benefit from information acquired by unions.

During the last eight years or so, more than twenty-five independent committees on occupational safety and health (COSH) have sprung up around the country. Reflecting labor's growing concern about occupational health hazards, COSH groups count among their members union activists, doctors and nurses specializing in occupational medicine, and industrial hygienists. Their primary goal has been to assist workers in improving the health and safety conditions in their workplaces, but several COSH groups have also exchanged information on toxic chemicals with local environmentalists. In 1982 the New York COSH, for example, organized a conference in which railroad workers who sprayed tracks on Long Island with herbicides met with people who lived near the railroad to discuss the health effects of such practices.

A community group seeking information on a hazardous material will find that a systematic check with any union representing workers who are engaged in the production, transportation, storage, use, or disposal of such substances will often yield valuable results. Not only can this association provide useful evidence linking the suspected hazard to health problems, but the communication between the environmental groups and labor unions can help to initiate long-term cooperation between the two.

Environmental Groups

National environmental organizations can provide local groups with several kinds of information. The larger ones maintain libraries with reference books and other resource materials. Their staffs are knowledgeable and can point to relevant sources of information. Several of the organizations publish books and reports. The Sierra Club, the Environmental Defense Fund, and the National Wildlife Federation distribute manuals on citizen action (see Resources for Action at the end of this book). The Conservation Foundation, another Washington-based group, issued an assessment of Reagan's record on environmental protection in 1982,[28] while a National Clean Air Coalition report on sources of air pollution in major cities around the country is an aid in targeting local polluters.[29] National environmental groups also employ scientists, lawyers, and other experts who can answer citizens' questions and help in developing a realistic plan for gathering information.

Since national environmental organizations are frequently in touch with local groups across the country, they often know who has useful data or similar experiences and problems. A Tennessee group fighting a toxic waste facility being built by SCA Corporation got the names of communities engaged in similar battles with that company from a national environmental organization. What they learned from these groups about SCA's operations in other places convinced them that their opposition was well founded.

Conferences sponsored by national groups are often a good place for activists to gather both technical and political information. A 1981 meeting sponsored by the Conservation Foundation and the Urban Environment Conference, and a 1982 get-together on hazardous waste sponsored by Ralph Nader, gave many leaders of local struggles an opportunity to learn new approaches to their communities' problems.

In some areas of the country local environmental groups have pooled resources to form state or regional coalitions, particularly for the purpose of sharing information. The Northwest Coalition Against Pesticide Misuse and Tennesseans Against Chemical Hazards are two coalitions that provide their members with up-to-date information and resource materials. Even in the absence of formal links, local groups share information. The New York City coalition against hazardous materials transportation regularly exchanged files and news clippings with the Greater Newark Bay Coalition Against Toxic Waste, which was also battling the Port Authority of New York–New Jersey.

Other National and Community Groups

Many other national groups also become involved in environmental health issues, including the League of Women Voters, various peace groups, public interest organizations such as Consumers' Union, the American Lung Association, safe-energy groups, citizen action groups such as the Association of Community Organizations for Reform Now (ACORN) and Fair Share. All are potential sources of information on both the technical and political dimensions of combatting an environmental hazard.

Tenant groups, neighborhood associations, and civic clubs can also provide a wealth of information. They often know about previous problems in the area, are familiar with the local power structure, and have learned through experience how to mobilize people. Neighborhood associations in Philadelphia helped to document the range of that city's toxic hazards for use in a right-to-know educational campaign. An antipesticide group in Kansas learned about the dangers of pesticides from a beekeepers' association. By tapping the knowledge of other community organizations, environmental activists can broaden their own understanding of their community.

How to Organize Information

As a group begins to investigate an environmental problem, information will begin to pour in. What should be done with it? How can members avoid being drowned in a sea of papers?

First, it makes sense to keep all the accumulated material in one place, so that everyone knows where to find it. A common library also helps to prevent time-wasting duplication of effort. If possible, choose a place that is easily accessible: a basement or garage, the office of a sympathetic community organization, or a church. Bonnie Hill, the leader of the Oregon group that won an EPA ban on herbicide spraying, had the novel idea of carrying files in the back of her station wagon, so they were readily available for meetings or public hearings.

Second, a good filing system is essential. While there is no magic formula, a system that organizes information so that it can answer any questions that are likely to come up is most helpful. Some groups are blessed with a member who loves to keep files; other may need to persuade someone to take on this task. Assigning responsibility to a single person or a small committee will ensure that one person will know where everything is.

Sometimes a group's library can become a valuable asset for other groups in the region or even across the country. One of the goals of Stop Project ELF, a Wisconsin-based group seeking to end a U.S. Navy Project that uses extremely low frequency (ELF) waves to communicate with submarines, is to become a repository for information on the health effects of such waves.

Third, the volunteer librarian should hold on to every piece of evidence. "Every little article I read I saved," says Edwina Cosgriff of BLAST. "It didn't always make sense then, but I figured they might fit together later."[30] What seems irrelevant at one stage of a struggle may become crucial later on. For example, due to the vagaries of court proceedings, the terrain of the battle to block Westway, the superhighway planned for lower Manhattan, shifted from mass transit and air pollution to the highway's impact on the Hudson River's striped bass population. Careful research and good files enabled Westway foes to take on the new issue immediately.

Finally it is important to know why the information is being collected. It may be useful to periodically review what has been learned. "What do we already know?" "What do we still need to know to advance our campaign?" Answering these questions helps to organize priorities for future research and to link information to a political strategy.

Gaining the Confidence to Do Research

For people encountering a suspected environmental hazard, gathering information is often the first step in trying to correct it. For some people, this is an intimidating task that can derail them before they start. But for those who can overcome initial self-doubt, success in becoming skilled researchers is in itself a satisfying accomplishment.

Activists explain their new confidence in different ways. Some attribute it to common sense. Verna Courtemanche, who discovered that a dump site near her home in Swartz Creek, Michigan, contained C-56, a component of the banned pesticides Kepone and Mirex, told a reporter: "I didn't need to be a professional to know it was a problem. It's like I don't need to be a mechanic to know that when a car hits you it will hurt you."[31] For Helene Brathwaite, who led the fight to rid her children's Harlem school of flaking asbestos, it was the realization that there were no real experts to turn to. "Nobody knew any more about this than I did," she said. "If you assume you're going to experts to help you, you're in trouble. Most of the time on environmental issues there are no experts, and if there were we wouldn't have these problems. So it's getting a body of knowledge for yourself when you decide you're going to attack this problem. Do your own research, so that you're well armed enough to go into a discussion of these issues with anybody."[32]

Those who have been through the process of collecting evidence, convincing others to join them and winning some victories can use their success to vindicate the long and frustrating research process. An activist from upstate New York whose group helped to force the cleanup of a toxic dump site put it this way:

The best strategy is to become an expert yourself on the problem you face. Our successes have come when we have had highly motivated people who are willing to really pursue an issue—learn it, ferret out materials from other agencies, find other groups who deal with the same agency or industry in other matters, and so on. No regulatory agency cares as much as you do about your problem with the urgency you feel. The best way is to get smarter than they are about your problem. This might seem difficult but most regulators are people, like you, who have had "on the job" training. Duplicate that in your spare time and you're halfway to victory!

5. BUILDING AN ORGANIZATION THAT CAN SURVIVE

Case History: We Who Care, Rutherford, New Jersey

Rutherford, New Jersey, is a comfortable suburb only eight miles west of New York City.[1] Many of its residents moved there to escape the more industrial landscape of the surrounding area. One such family was the Cleffis. For Vivian Cleffi, a housewife, and her husband, Jim, a trucker, Rutherford seemed like the ideal place to raise their three sons.

Then in 1975 their oldest son, eight-year-old Jimmy, developed a rare form of leukemia, a cancer of the blood. In the waiting room of the New York City cancer specialist who was treating Jimmy, Virginia Cleffi met two other Rutherford parents whose children had leukemia. Alarmed by what she felt was an unlikely coincidence, Cleffi began to suspect that something in the town was making its children sick. With the help of the Parent-Teacher Association (PTA) at her children's school, she started to look for other cases. She found eleven cases of leukemia and Hodgkin's disease, a related cancer of the lymph system. Six of the victims were enrolled in or had attended the same elementary school.

At first the parents were reluctant to act. As one mother put it, "We were kind of afraid of saying anything that was going to scare everybody, and maybe be off base, maybe it was just something in our imaginations that it was a problem."[2] But the reluctance of the state department of health to investigate promptly angered the parents. And growing media coverage, as well as rumors that the illness might be contagious, led them to think that they had to get accurate information to the public. By early 1978 two of the children, including Jimmy Cleffi, had died.

It was at this point that the parents organized a group called We Who Care. Its purpose, according to Carol Froelich, a mother and founding member, was to "share information and make sure that everything that could be done would be done and that the information the state found would be shared with us."[3] We Who Care's membership included parents, PTA officials, and representatives of senior citizens centers, churches, and women's clubs. The group asked state officials to attend a meeting to hear their complaints. Seven hundred people came. The mother of one of the children who had leukemia explained to a reporter what the residents wanted: "They want the answers, they want the state to find out, they want everything undertaken that's feasible."[4] An epidemiologist from the state department of health informed the meeting that Rutherford's leukemia rate was six times higher than the national average.

Members of We Who Care began to look for the cause of the cluster of cancer cases. Their research did not reassure them. Forty-two industrial concerns using organic chemicals—many of them carcinogens—were within three miles of the school. Two sides of the town are bounded by the Hackensack Meadowlands, a swamp that is sprayed for mosquitoes with more than 50,000 pounds of insecticide every year. Until it was banned in 1967, DDT, a carcinogen, had been used in mosquito control. Two major highways flank the town, showering it with automobile exhaust, another suspected carcinogen. Finally, Rutherford is situated at the convergence of three microwave beams, a form of radiation that may cause cancer, from two airports and an industrial research facility.

Faced with an organized community group and a barrage of media coverage, the New Jersey Departments of Health and Environmental Protection investigated the outbreak with an epidemiological survey and air, water, and soil sampling. Despite these efforts the state was never able to find a specific cause for the cancer cluster.[5] While this failure disappointed and frustrated residents, We Who Care did not disappear. It brought in its own researchers to conduct further studies. It critiqued the official investigation and forced the state to make futher tests. It organized a telephone campaign that deluged officials with calls whenever a chemical odor appeared. The group also forced the town government to shut down a nearby plant that used benzene, a known cause of leukemia, in violation of environmental regulations.

These accomplishments, however, failed to quiet a controversy that was surfacing within We Who Care. Some members of the group, including several parents of leukemia victims, wanted to take a more militant course. They felt they had enough information to take action against certain local polluting industries and believed the state was not conducting a thorough investigation. They also accused local politicians of failing to put pressure on the state to act.

The other faction wanted to follow a more moderate line. They were satisfied that the state had made a genuine effort to investigate; their strategy was to work with elected officials to lobby for changes in regulations and enforcement. Carol Froelich explained how this faction felt about the more militant members: "I don't think they gave us a very good name and after a while some people in town were thinking we were kind of crazy. I think a few people stopped being active in the group because they didn't want to be associated with a group that was always accusing others of not being truthful and of not being satisfied."[6]

In order to pursue what they felt was the correct direction, the moderates wrote organizational by-laws that mandated a new election of officers. The more militant leaders were defeated and soon became less active in the organization.

We Who Care became smaller, with more diverse interests and a focus on education and lobbying. These efforts achieved some long-term results. With the support of labor and environmental groups, the group forced New Jersey to establish a cancer registry, an important tool for future epidemiological research. One member was appointed to the Rutherford Board of Health, which is now more actively involved in environmental health issues. Local newspapers now cover environmental stories regularly, often turning to We Who Care for quotes or information, and the local PTA developed a cancer-prevention educational program that has served as a model for schools around the country.

Despite these advances, however, the fear has not left Rutherford. New cases of leukemia and Hodgkin's disease are diagnosed periodically. As Vivian Cleffi observed, "You look at the kids and here they are, they're playing in the same rotten air. You look at them and you can't help thinking to yourself, which one of them is going to be next?"[7]

Once residents have uncovered a threat to their health and learned what they can do about it, their next step is to take action to protect themselves. Most people discover very quickly that they can do very little as individuals. One-woman or one-man battles take too much time and energy and yield too little satisfaction and too few successes to sustain anyone for very long. Thus most environmental activists find they need to organize or join a group that can gain the experience, resources, skills, and, most importantly, the political power necessary to combat corporate or government polluters.

Such groups begin in a variety of ways. Most often a few people worried about an environmental problem begin to talk to each other. "Our group got started with a few concerned people meeting in a friend's kitchen," reported one Midwestern activist. Two individuals who had been upset by foul odors emanating from a factory in Brooklyn, New York, went to the factory manager with questions. When he gave them the runaround, they decided to form a committee to investigate the problem. A Wisconsin group seeking to ban uranium mining began when several people decided to pool their efforts and skills so as to broaden their support.

In other cases a few people will decide to call a public meeting. The Oregon organization opposed to herbicide spraying got started at a meeting called by several worried neighbors. Residents of several towns in Minnesota organized a public information meeting on the health effects of uranium mining. More than six hundred people attended and an organization was formed.

Sometimes those who start a group will have had previous experience in organizing. Environmentalists in Broward County, Florida, asked everyone who had worked in earlier local campaigns to join a new organization to fight water pollution. Several individuals in Hawaii who were affiliated with the national Sierra Club got together to combat growing air and water pollution. But most groups are formed by those without previous political experience. Lois Gibbs of Love Canal and Nell Grantham of Hardeman County, Tennessee, are both examples of women without a history of activism whose family and community responsibilities led them to organize environmental campaigns.

Still another way for a group to start is for an existing organization to expand its focus. A food cooperative in Idaho, for example, became concerned about contamination of its produce and began to fight the unnecessary use of pesticides and herbicides. In recent years local chapters of citizen action groups such as ACORN (Association of Community Organizations for Reform Now), Mas-

TABLE 2 Advantages and Disadvantages of Various Organizational Forms

Type of Group and Examples	Advantages	Disadvantages
Concerned citizens' group: Love Canal Homeowners' Assn. Hardeman County, Tenn. We Who Care BLAST	1. Purpose clear 2. Members' commitment high because they are affected by problems 3. Flexible in structure, strategy, tactics 4. Moral credibility high	1. Difficult to find needed resources 2. Not connected to regional or national networks 3. Tend to have short life spans, making monitoring of victories difficult 4. May have parochial perspective
Community coalition: Washington Heights Coalition Against Hazardous Materials Transportation Coalition for United Elizabeth Greater Newark Bay Coalition Against Toxic Waste	1. Represents broad spectrum of community 2. Access to community resources (e.g., office, volunteers) 3. Brings new constituencies into environmental issues	1. Members of coalition can have different agendas or concerns 2. Decision making can be slow 3. Organizations' commitment to coalition may vary or waver
Regional/national environmental group: Local chapters of Sierra Club National Audubon Society	1. Access to information and technical expertise 2. Credibility with legislators or policy-makers 3. Can advocate national policy	1. Some organizations have elitist image 2. May not attract labor minority support 3. Can be viewed as one more special-interest group 4. National group may compromise or manipulate local chapter
Multi-issue citizen action group: WARN Mass Fair Share ACORN	1. Community organization exists 2. Organizing experience and skills available 3. Political and financial resources exist 4. Can monitor victories 5. Can advocate national regional policy	1. May lose interest in environmental issue in favor of other issues 2. More interested in getting power than winning specific fights 3. Commitment to process of organizing can slow progress

sachusetts Fair Share, or Ralph Nader's Citizen's Alliance, previously concerned with housing, taxes, and utility rates, have turned to environmental health issues.

Choosing an Organizational Form

Whatever first brings people together, they must soon decide on the kind of organization they want to create. On Staten Island and at Love Canal, residents chose to organize concerned citizens groups, associations of neighbors seeking to remedy a specific hazard. The Washington Heights coalition against hazardous materials transportation and the Greater Newark Bay Coalition Against Toxic Waste chose instead to bring together several community organizations in order to build a more broadly based front. Activists in WARN and in Massachusetts Fair Share worked on several issues, including environmental health. These multi-issue groups often have chapters and a statewide or even a national political focus. Still another organization form is affiliation with a national environmental group such as the Sierra Club or the Audubon Society. Table 1 summarizes the advantages and disadvantages of each of these types of organization.

While there is no magic formula for choosing the ideal organizational form, activists can make more informed decisions by considering their goals and resources, and the opportunities that are present in their communities. An awareness of the weaknesses of each form makes it possible to foresee some of the problems that may come up, as well as ways in which to compensate for or even prevent them. For example, a concerned citizens group newly formed to block a toxic waste site might balance its lack of experience with an informal relationship with a national group specializing in this issue.

The case histories in this book demonstrate that organizations of all four types have won significant victories. Furthermore, many groups combine elements of several forms. The Coalition for a United Elizabeth, described in chapter 1, for example, is a citizen action group that addresses such issues as housing, education, and racism as well as toxic chemicals, but it is also a coalition of community organizations, including some concerned-citizens groups.

Building an Organization People Want to Join

Most local environmental conflicts are fought not by professional activists but by men and women who have jobs, families,
outside interests, and financial responsibilities. If busy people are
to put in the weeks, months, or years of unpaid labor necessary to
win these battles, they need to be part of organizations that work
efficiently and that give their members a sense of self-respect and
accomplishment.

Building an organization that people want to join—and stay
in—is as crucial to victory as careful research, a well-thought-out
strategy, and imaginative tactics. Community struggles can be
frightening or frustrating. BLAST activists, for example, had their
finances and personal lives investigated by the company that
owned the LNG tanks.[8] Members of a Pennsylvania group fighting
a toxic waste site were warned to desist; when they refused their
homes were broken into and vandalized. And of course many
people fear that their lives are in danger from chemical exposure.
Added to these anxieties may be industry's refusal to acknowledge a problem, government officials' indifference, and regulatory
agencies' painstakingly slow response. Only an organization that
helps people get through these situations can hope to win their
allegiance.

Environmental groups face four key organizational questions:
How does a group make decisions? What is the role of its leaders?
How is its work allocated? How are its goals chosen? While there
is no one answer to these questions, experience shows that the
choice can determine how loyal a group's members are, how
much they are willing to contribute, and how satisfied they feel
about their efforts. By thinking about these questions from the
start, organizers can anticipate problems and take steps to prevent
them.

The most common—and the most successful—way a community environmental organization makes decisions is through informal discussions leading to consensus. The leader of an
antipesticide organization in California reported that her group's
decision-making process was "very loosely structured. We decide
we need to get together, we decide what needs to be done, then
divide up jobs and do them." A member of another antipesticide
group, this one in Oregon, explained some of the requirements for

reaching decisions by consensus: "[We have] a lot of mutual trust. We all give input and listen to one another. Differences are respected. We tend to trust that those of us who have the energy to take action at any given time will do well for our common cause."

Reaching decisions by consensus can be time-consuming. The group must agree on a definition of the problem and generate a list of every possible solution. It must then review, change, and consolidate its proposals, and set priorities according to some mutually agreed-upon criteria. Finally, it must decide who will do what, when, and how.

What are the advantages of this cumbersome procedure? For one thing, it gives all members the opportunity to voice an opinion—nothing makes people feel they "own" a decision as much as having a say in it. Moreover, the consensus process gives the group the benefit of each person's special knowledge or perspective. Ten heads are more likely than one to foresee a problem and come up with a good solution.

Making decisions by consensus also strengthens the organization. In those groups where all decisions are made by a few leaders, their departure literally decapitates the organization. Conversely, when all members participate actively in making decisions, no one person is indispensable. In addition, the consensus method teaches new members about how the organization works.

Finally, working by consensus can help to prevent a group from splitting into warring factions. Robert's Rules of Order or other formal voting mechanisms force group members to choose one side over the other. Every decision creates winners and losers. For example, disagreements about whether to confront or cooperate with New Jersey state authorities divided We Who Care into two camps. The bitterness between the two eventually led the more militant members to become less active, thus depriving the group of their energy and commitment. Had the organization developed a method for hammering out a mutually acceptable strategy, it might have transcended the differences and grown stronger. While the consensus process does not discourage conflict, it depersonalizes disagreement. The ultimate goal of debate is to find a position that everyone feels comfortable with.

Taking the time to make decisions by consensus is probably one of the most powerful methods to keep a group's members active and involved. Nevertheless, the process has its drawbacks. In a crisis situation—a nuclear power plant leak, for example, or the imminent explosion of a truck carrying flammable liquids—it can take too long to decide how to respond. Furthermore, members with a genuine difference of opinion, as well as those seeking to sabotage an organization, can use the consensus process to prevent any decision from being made at all. As a result, those who joined in order to do something become frustrated and leave. Working people with little time to spare or those uncomfortable with long intellectual discussions are the most likely dropouts and their departure can isolate a group from politically important constituencies. Some groups' preoccupation with the process of making decisions dooms them to endless meetings that lead nowhere—here the ideal of consensus becomes a justification for doing nothing.

Fortunately, there are ways to counter these problems. For instance, to prevent disagreements from blocking a decision to act organizers of the Clamshell Alliance, the antinuclear group fighting the Seabrook nuclear power plant in New Hampshire, came up with a hierarchy of alternative ways to register objections to a proposal. These included:

Nonsupport   ("I don't see the need for this but I'll go along")

Reservations   ("I think this may be a mistake but I can live with it")

Standing aside   ("I personally can't do this but I won't block others from doing it.")

Withdrawing from the group[9]

Each offers a member who disagrees a respectful way of voicing his or her opinion without stopping a decision.

Some organizations divide the decision-making process in order to speed it up. Not every member participates in settling every issue. For example, a Pennsylvania group seeking to clean up water pollution emanating from a county landfill has an executive committee that makes day-to-day decisions, while major decisions about policy and direction are made by the group as a whole. Several groups have developed guidelines on how to respond to an emergency and empowered leaders or a committee to make on-the-spot decisions. Other groups decide on a time limit for a discussion before it begins.

By using these methods, by incorporating consensus decision making into every level of an organization and by adhering to it regularly, some groups have learned how to make quick decisions even with large numbers of people. On November 15 and 16, 1981, 3,500 women from around the country gathered in Washington, D.C., to demonstrate against Pentagon military and environmental policies. Many of the decisions about what to do and how to do it were made on the spot by all the participants. The local planning meetings for the action had given the demonstrators a framework for achieving consensus rapidly.[10] Similarly, the Clamshell Alliance was able to use participatory decision making when 2,000 people occupied the Seabrook nuclear power plant in 1977. Even after the arrest of 1,414 of the demonstrators, decisions continued to be made collectively. One of the participants described how the Clamshell's decision-making process affected him:

> There's a whole list of things that I got out of the Clamshell structure. First of all I got a sense of security. I knew that no decision would be made unless I was consulted. That's important to know when you're about to be arrested for a nonviolent act. Secondly, it built a sense of community. My affinity group [a few people who stayed together throughout the occupation] became my ad hoc family, and we made deep friendships going through the many crises together. Also, I felt ownership—or maybe partnership—in the occupation. I felt responsible for how well it went. The training that the structure had set aside ahead of time prepared us as groups capable of making decisions. This put power in everybody's hands. We had to think through consequences beforehand.[11]

Another way for groups to make decisions is to elect leaders and then delegate decision making to them. BLAST and We Who Care, for example, depended on the dedication and judgment of a handful of people who put in hours of time to keep these organizations going. The advantage of this method is that it is efficient—the few activists get to know each other well and, with experience, can accomplish their tasks quickly. The sense of accomplishment that such efficiency engenders helps to keep the participants' level of commitment high. For small groups whose members already know and trust each other, or for battles that can be won rapidly, this method of making decisions can work well.

But in the long run there is a price to pay for such efficiency.

Those who work so hard find it more and more difficult to include new people. They become discouraged when no one else seems as committed as they are. BLAST's Edwina Cosgriff labeled the group of four who did the most work as the "George group" since other members always seemed to say, "I'm too busy, let George do it."[12] The "Georges" soon become resentful and tired and eventually they may burn out, then drop out. The short life span of many concerned citizens groups is evidence of their failure to integrate members into the decision-making process.

A second factor influencing how members feel about their group is the role of its leaders. Leadership can be defined in different ways. "My only 'official' function is to activate our phone chain for meetings," reported the president of an upstate New York environmental group. On the other hand, the leaders of a North Carolina organization were responsible for "organizing all the activities of the group." Most leaders define a role for themselves somewhere between these two extremes. Among the common responsibilities of leaders of environmental groups are coordinating and organizing activities, serving as spokespersons to the media and other organizations, running meetings, educating the membership, and devising strategies. In order to broaden their leadership, many groups divide these tasks among individuals or committees. Others rotate leadership on a regular basis.

The style of leadership strongly influences the participation of members. Authoritarian leaders often create passive or hostile members. If the president is going to make all the decisions and then tell you what to do, why think for yourself? The flip side of this coin is that weak, passive leadership often breeds frustration and cynicism. It is a common mistake to think that strong leadership and democracy are contradictory. Strong leaders are a model for others—they inspire them, challenge them to think, and force the organization to define problems and search for solutions. Strong leaders can help a group to create ongoing democratic processes and structures.

One of the most effective ways an organization can build strong leadership is to recognize and value different types of leaders. The person who can organize and carry out a leaflet campaign within twenty-four hours, the one who has the courage to argue publicly with the senator who promises the moon but delivers nothing, the one who puts on a successful fund-raising event, the first one who volunteers for civil disobedience—all are leaders who can move the group forward. At different stages in an organization's life one style of leadership may predominate, but if alternative styles and leaders wither from lack of practice or recognition, the organzation as a whole will suffer.

Leaders of community organizations encounter a variety of problems. Anticipating them, and planning strategies to counter them, may prevent the individual headaches and organizational strife that can make community organizing so frustrating. One common problem is challenges to the leaders by members of the group. Some will respond by ignoring the challenge, hoping it will go away. This rarely works; it also misses an opportunity to help the organization grow. Most dissent arises not from personality conflict but from political differences. By acknowledging these and encouraging debate, leaders can help a group engage in mature political discussion. And if a leader candidly admits to differences, the dissident group's attempt to portray the leaders as undemocratic or unresponsive will be undermined. To return to the example of We Who Care, a frank discussion of the political differences between the two factions might have helped both sides define their strategies and goals more clearly. Instead, the new leadership attributed the original leaders' desire for a confrontation to negativity and bitterness and excluded them from the group's inner circle.

If, on the other hand, disagreements are aired, the group can then decide how to proceed. A member of an upstate New York toxic waste group explained how environmentalists in her community decided to handle their differences: "We have a 'sister group' nearby that . . . has a totally different approach to environmental affairs. We tend to be legal, academic, while they are radical and love demonstrations. It's a nice complementary situation, since both types rarely coexist in the same group for very long."

In another example, the Love Canal Homeowners' Association (LCHA) chose to maintain an uneasy relationship with a dissident faction rather than pursue an amicable divorce. The subgroup, which called itself the Action Group, criticized Lois Gibbs and others for not demanding government action forcefully enough. The Action Group organized demonstrations, issued leaflets, and spoke at public meetings. But because its members were local residents who had been active from the start, because the LCHA wanted to maintain a united front, because the Action Group's more militant demands made the larger group seem more reasonable, and because Gibbs feared a breakaway group might become violent, the LCHA leadership decided to opt for coexistence.[13]

Adeline Levine, a sociologist who studied the Love Canal disaster, described some of the benefits of this uneasy relationship:

> [The LCHA leaders] learned that, difficult as it was, they could live with the criticism. They learned to evaluate the criticism, separating out the style of criticism from the substance and adopting and adapting the creative suggestions. They also learned how to control the opposition group through their own creative use of the rules, and they learned when to go along, when to compromise, when to assert control openly and firmly, and when to call bluffs by calling for votes of confidence.[14]

Some groups have avoided leadership challenges by excluding certain people. Edwina Cosgriff of BLAST reported that her group "had gotten rid of all the political animals who came to meetings just to push one candidate or another."[15] Other groups have limited membership to those who live in the community. The danger is that, in their zeal to avoid a "radical" image, a group will exclude anyone who disagrees or anyone with a background in political activism. This can hurt in two ways. First, these so-called radicals are often experienced, sophisticated activists who can contribute to planning a successful strategy. Second, their participation in political discussion forces a group to examine its own premises and strategies, which often helps clarify its choices. Another tactic for managing a challenge is to set rules for choosing new leaders. An established process for endorsing or rejecting leaders allows a group to discuss what it wants from its officers without disrupting the organization.

A second problem is how to develop new leadership. On the one hand, leaders may come to like power and find it difficult to share the limelight. On the other hand, they may become tired and want to find new people to share the work. There are several ways to bring new people into leadership positions. An orientation session where new members learn the history of the organization, its structure, and the issues it has confronted not only prepares new participants to become effective spokespeople for the group but also saves time at regular meetings, since basic questions have already been answered. Energetic new members can be given some leadership responsibilities as soon as they join. The best way to learn how to be a leader is to be one. Pairing experienced and inexperienced members helps to develop leadership skills. And since those who are directly affected by an issue often make good leaders, bringing them into the organization adds to its leadership potential.

Effective leaders respond to suggestions and pressure from their constituents. By creating structures whereby future leaders have to take other members' opinions into consideration, an organization trains its leaders in the democratic process. It also helps create a situation where new leaders emerge from the membership rather than being selected by those in authority. Subdividing a group geographically (by neighborhood or block) or by task also helps new leaders rise from the ranks.

Even when the leadership is committed to creating new leaders, the group may not have members who want to take on new responsibilities. The resolution of this problem requires a careful diagnosis of its cause. Sometimes members are intimidated by those in charge; the leaders may seem so educated, articulate, and skilled that no one can equal them. Leadership training programs can help to familiarize members with the information and skills they need. Demystifying the tasks of leadership makes them seem less intimidating. National training centers such as the Midwest Academy in Chicago offer courses in leadership development.

In other situations the leaders may unintentionally create a closed clique of those in positions of responsibility. Since those starting a group often turn first to their friends and families for assistance, an outsider who wants to participate may feel excluded. This is especially true if the initiators are of a different class or race than most of the community residents. The Washington Heights coalition against hazardous materials transportation, for example, was formed by white professionals. At first Hispanics were reluctant to take on leadership roles. Only when the original leaders were absent for a few weeks, leaving the group to plan a major action, did Hispanic residents emerge as leaders.[16] Sharing work, socializing, and holding frank discussions of the cliques that do form can help to break down barriers.

Finally, unrealistic expectations can discourage future leaders from taking on new responsibilities. If members have to choose between being a round-the-clock volunteer and doing nothing, they will often opt for the latter. Offering potential leaders a series of responsibilities with differing time and energy requirements allows a greater choice in deciding what level of commitment to sustain.

The third factor that affects the way in which members participate is how the work gets done. It is all too common for a few activists to do most of the early work in an environmental campaign. New people who join are often less knowledgable and experienced, and so are given routine work: typing, answering phones, handing out leaflets. As the group gains recognition the founders inevitably talk to the press and television and speak at public meetings—after all, they can do it best and they "deserve" the recognition for their hard work. After a while the typists and telephone answerers stop coming around. They begin to suspect the motives of the leaders or doubt their own ability to make a meaningful contribution. Soon the organization falls apart.

Can this process be prevented? Is there another way to divide up work that draws people into an organization rather than driving them out? Several alternative methods have been tried. First, successful groups have outlined a number of activities in which its members can participate. Not every community resident has the ability to devote full time to an environmental cause and for the more devoted to expect the same from all can only mean constant disappointment. A victorious campaign needs people to circulate petitions, write letters to politicians, attend demonstrations, make telephone calls, edit flyers, make posters, speak at public meetings, collect funds, plan strategy, read scientific reports, and so on. No matter how little time a supporter has, there is always something to do. Participating in community affairs can be addictive—the person who canvasses his or her block this week may want to write a leaflet next week to answer the questions that neighbors asked. But whether a new member volunteers increasing amounts of time or not, the organization that recognizes and values every kind of participation is the one that is likely to attract and hold newcomers.

Second, successful organizations tap their members' skills. Those with scientific training evaluate technical reports; older residents help to track down what was dumped where twenty years ago; parents survey their children's health problems. When new members are asked to do what they can already do well or easily, they will volunteer more readily. But people also want to learn new skills. A strong training program is an excellent insurance policy against individual dropout or organizational death. Sharing skills can strengthen the group and eliminate the need to learn new skills from scratch. Some organizations delegate a few members to learn a new skill (e.g., how to do a health survey, or how to prepare a newsletter); they then teach it to the group as a whole.

A committee structure can also help members learn new skills. The research committee, for example, or the media committee can include one experienced person and a few neophytes. Eventually every task has several competent practitioners. A variation on this structure is to pair an experienced person with a new member. The Coalition for a United Elizabeth, for example, always sent two people to represent it at meetings. At first the experienced person would speak and the newcomer would listen. With time, however, the roles were reversed. The task of the old-timer was to listen to the more recent member, then make suggestions and offer constructive criticism. Soon the organization had a stable of experienced public speakers.

Any organization has a certain amount of routine, boring work, and how that work is allocated has a profound effect on its members' attitudes. Having the "stars" do all the exciting work and the others do all the drudgery invariably leads to bad feelings. The work that requires no special skills—stuffing envelopes, putting up posters, etc.—should be done by everybody. A work party to do a mailing can be made into a social event, and old and new members can get to know each other. It also gives the stars a more realistic view of what it takes to keep an organization going.

For routine work that is more skilled—typing, mimeographing, bookkeeping, answering the phone—there are several ways to share the burden. A skill can be taught to several members, or those who type or answer phones can also be included in more glamorous tasks. And the importance of these skills to the organization should be publicly acknowledged: a good leaflet needs a skilled typist as much as a competent writer.

It is particularly important for a group's female members that the routine work be divided fairly. Too often women do the typing, filing, and telephoning while men take on the more exciting jobs of planning strategy and talking to the public. Not only does this division of labor deprive a group of the wisdom of half its members, but it also increases the likelihood that women will drop out. Who wants to volunteer for a group that treats you like a second-class citizen?

Some community groups succeed in dividing up the work when they have only a few members but develop problems as they grow. It is, after all, much easier for ten people to work collectively than for one hundred. To prevent growth from leading to too much specialization and bureaucratization, some groups have experimented with decentralization. The Love Canal Homeowners' Association, for example, broke into block committees. Members living on a particular block were responsible for planning and carrying out the association's work on their street. Like an amoeba, some of the antipesticide groups in the Pacific Northwest divided as they grew. Decentralization allows more people to assume leadership and take on planning responsibilities. If necessary, a committee can coordinate the joint activities of the new groups.

Before a new organization can begin a campaign to protect its community, it faces one additional crucial task: it must define its goals. A goal is a statement of direction and purpose, specifying what the group wants to accomplish, and choosing it is perhaps the most important decision an organization will make in its early stages. Only by knowing *what* its members want to do can a group decide *how* to do it—a goal is thus the starting point for planning a strategy or choosing tactics. Having a clear-cut goal also makes it easier to set priorities and allocate limited resources. How much time should be spent on lobbying senators? On conducting a health survey? On recruiting new members? By understanding the relation of these activities to the ultimate objective, it is easier to decide which is most important. Finally, a goal can provide inspiration, reminding people why they are spending so much time and energy on the project. In moments of despair, frustration, or internal conflict, people need to remember their original motivation: to protect their children's health, to prevent a disaster, or to create a livable community.

The following examples illustrate some of the goals chosen by environmental groups:

—To force Reserve Mining Corporation to discontinue dumping tailings into Lake Superior.

—To eliminate the use of all toxic chemicals for pest control within our watershed area; to support the faltering economy by providing more labor-intensive forestry through viable manual alternatives.

—To seek removal of PCBs stored in high-density residential areas; to seek adequate and safe storage of PCBs in these areas pending ultimate disposal or removal.

Each of these statements clearly specifies what the group hopes to accomplish.

Choosing goals is always a *political* task. It determines who will be friend and foe and what will constitute success or failure. Thus the choice made must be based on an analysis of the causes and likely solutions to the problem. Among the questions a group must consider are:

—Does it seek a solution at the local, regional, or national level?

—Who should be responsible for remedying the hazard?

—How can the goal be defined so as to maximize support?

—Does the goal include an alternative to the hazard (for example, hand clearing as an alternative to herbicide spraying)?

—How should a specific goal (closing down a dump site) be balanced against a broader and more general one (building a strong environmental movement)?

By consulting the sources of information described in the previous chapter and by discussing the above questions, a group will be better able to select goals that will shape its strategy.

People join environmental groups for a variety of reasons: concern about their families' health, a desire to protect their community, a commitment to social justice. But what keeps them active is the feeling that they as *individuals* are making a difference, learning new skills, and contributing to a better world. The statements of concerned residents about what they have gained from their experience shows this again and again.

"I was quite shy before, but now I guess I'm an extrovert," said an activist from Love Canal.[17] And a woman involved in a fight against toxic contamination of drinking water in a small town in

New Hampshire told a reporter, "When it comes home to roost, you'd be surprised how quickly private citizens can become adept at pressuring Congressmen, doing complex research, making calls and generally raising hell to protect your family and your town."[18] An organizer of the Harlem school asbestos fight put it this way: "I'm as competent in the problem-solving area as the experts are. If a problem comes up again, I know exactly what to do."[19]

This sense of pride and accomplishment comes from winning concrete victories and from the feeling that lives have been changed for the better. Not every organization can expect to win all its battles but by creating a group that makes its members feel more competent, more powerful, and more connected to others, it achieves something equally important: it helps to nurture individuals who in the long run will have the courage, compassion, and self-esteem to fight for a society where well-being and human need take precedence over profit and greed.

6. STRATEGIES FOR ACTION:
Community Education

Case History: Delaware Valley Toxics Coalition, Philadelphia, Pennsylvania[1]

On January 22, 1981, the Philadelphia City Council passed the nation's first municipal right-to-know bill.[1] The law forced a manufacturer to disclose both to the public and to its workers which toxic chemicals are produced, stored, transported, and emitted by each of its factories. Robert Vogel, chief regulatory counsel to the Rohm and Hass Company and the industry representative in negotiations on the bill, called the new law "a reasonable, businesslike, and cost-effective piece of legislation."[2] Consumer advocate Ralph Nader called the law "a model for municipalities throughout America."[3] Abandoning his earlier opposition, Philadelphia Mayor William Green signed the bill. An activist involved in the campaign explained the broader vision behind the law: "We want the community to be in control of toxics in their neighborhood. Through right-to-know we can get information to make informed decisions on what we will allow in our neighborhood and under what conditions the companies will operate."

How did this victory come about? How did environmentalists and labor unions join together to persuade the city council to pass this precedent-setting legislation and how was industry convinced to accept it?

Philadelphia's right-to-know bill was the culmination of years of educational work. Since 1975 the Philadelphia Project on Occupational Safety and Health (PhilaPOSH), an organization of union activists and health professionals, had been working to identify and correct occupational hazards, especially exposure to toxic chemicals and carcinogens in the many chemical plants in the area. In 1979 PhilaPOSH and the Environmental Cancer Prevention Center, a community education program of the Philadel-

133

phia Public Interest Law Center, jointly organized a citywide conference on chemical killers. Over 350 neighborhood, labor, and environmental activists gathered to learn about toxic chemicals in the community and the workplace. Community organizers had become interested in environmental issues following battles about asbestos in the schools, a warehouse storing polychlorinated biphenyls (PCBs) in a residential neighborhood, and episodes of air pollution-related illnesses.

Out of this conference emerged the Delaware Valley Toxics Coalition (DVTC), the organization that coordinated the right-to-know campaign. The group used a variety of tactics to make its case. At one city council hearing, for instance, a member of the United Automobile Workers' Union brought in a canister of carbon dioxide labeled only with a code number. As he turned on the valve, spraying the hearing room with an invisible gas, city council members shouted, "What's in that stuff? Shut it off!"[4] "This is what we face every day on the job," he replied. A mother whose children had respiratory illnesses from air pollution also testified. Personal stories made good headlines, while testimony from medical and scientific experts helped to convince those who needed more stringent evidence. Caron Chess, a leader of the DVTC, explained how the hearings educated the public about the issues: "There was a whole lot of education going on at those hearings: education of the city council by those who attended, as well as among the community and labor people involved." Jim Moran of PhilaPOSH added, "If we had passed right to know in the first thirty days of having raised it, I don't think it would have been for the best. By working to get support for right to know passage, we ended up educating the public. You mention right to know and toxics to people and they respond, 'Oh yeah, I heard about that.' If right to know had just slipped through, nobody would really have understood it."[5]

Since the campaign required work at many levels, everyone who wanted to contribute was able to. Chess explained that the right-to-know campaign "gave people something to do about toxics, from low level involvement—signing a postcard to a council person—up to testifying at a hearing and organizing others to do the same, as well as petitions, phone call drives, letter writing campaigns, and mass lobbying of council people by large groups of people."[6]

As a result of these efforts, the DVTC built a broad spectrum of support. Firefighters who had to extinguish toxic chemical fires endorsed the bill, as did the League of Women Voters, Vietnam veterans who had been victims of Agent Orange, senior citizens organizations, and neighborhood associations.

At first industry opposed the bill, arguing that it was overly complicated and expensive. But strong political mobilization and favorable media coverage helped to convince manufacturers that the bill could not be defeated. The coalition also made some compromises. Industry wanted a trade secrets exemption, giving them blanket rights to deny information on specific products. The final bill included a narrower exemption, one that gave doctors the right to receive all information, trade secret or not.

Passage of the right-to-know bill marked the end of one stage of a struggle and the beginning of another. The coalition must now ensure that funds are appropriated for its enactment; that regulatory agencies are monitored to see that they enforce the law; and that the public is educated about how to use the information the law provides.

The campaign has already had beneficial effects both in Philadelphia and around the country. In 1983, for example, the Philadelphia Chapter of the Association of Community Organizations for Reform Now (ACORN) successfully blocked construction of a toxic waste incineration facility in a crowded low-income neighborhood; their organizing was successful in part because of the awareness and concern about toxics generated by the DVTC. And by the end of 1983 at least thirteen states and thirty-five cities had passed right-to-know bills,[7] often at the insistence of activists using educational materials and strategies developed in Philadelphia.

You have discovered an environmental hazard that is threatening your community and you have researched its causes thoroughly. You and some neighbors have organized a group to combat it. The next task is to decide what to do. The following three chapters will describe four methods that have been used to protect different communities: public education, legal action, legislative advocacy, and community organization.

A community education campaign is often the first step in

mobilizing public opinion against an environmental threat to health. The Delaware Valley Toxic Coalition's public education campaign undoubtedly laid the foundation for the passage of the right-to-know bill. Without the community forums, media coverage, flyers, and testimony at hearings, the city council would never have been convinced to pass the legislation. Community education has three goals: to inform people of existing or potential threats to health, to convince people to act, and to demonstrate the need for ongoing organizations that can fight in the political arena.

Planning a campaign to realize these goals can be divided into five steps. The first is to assess the concerns of community residents. Effective learning—learning that makes people act differently—is most likely when the learners' real concerns are addressed. Thus if those fighting a nuclear power plant talk only about high utility rates and big business's control of the economy, while residents are really concerned about whether they will get cancer, the educational campaign may well fail to win adherents.

Community groups have used several techniques to solicit their neighbors' views. Lois Gibbs and other members of the Love Canal Homeowners' Association (LCHA) knocked on every door, asking people whether they had health problems. This not only gave the LCHA direct knowledge of its constituents' concerns but also established a relationship between activists and residents. Other groups use door-to-door canvassing to poll residents for their opinions on environmental legislation, to survey common health problems, and to solicit views on the major problems facing the community. Canvassing permits a dialogue between canvasser and resident. The canvasser can learn what people are thinking and provide them with information about the suspected hazard. It is also a way to identify potential supporters. Some groups use a written questionnaire; others prefer a more open-ended, less structured format.

People's concerns can also be learned through community speak-outs, by mail surveys, or by interviewing residents who are knowledgeable about their neighbors' views (e.g., the president of the Parent-Teacher Association [PTA], the leader of a religious organization, or a trade-union activist). A review of previous struggles in the community may show what moves people to action. Activists in a New York City neighborhood decided to focus their campaign against childhood lead poisoning on poor housing conditions (which contribute to poisoning) after they learned of previous rent strikes, housing coalitions, and similar tenant actions.

The second step is to choose the few key messages that will form the core of the community education campaign. Two criteria determine the choice of messages. On the one hand, educators want to alert people to the most serious dangers the hazard poses; this is done by analyzing and editing previous research (collected as described in chapter 4). On the other hand, they want to present information that will address the neighborhood concerns solicited in the needs assessment phase. For example, the Delaware Valley Toxics Coalition educated people both about problems already worrying some neighborhoods, such as asbestos in schools, air pollution, and PCB storage, and also about potential dangers, such as contamination of the ground water by carcinogens and accidents involving the transportation of hazardous materials. Since community education is least effective in communicating complicated information, it is important to choose a few specific messages.

In the third step, organizers target subgroups for specific information, dividing the population according to their concerns. For example, parents may be especially interested in their children's health. Homeowners may be more concerned about property values than are those who rent. Members of religious groups may respond to moral arguments. Factory workers want to know how the hazard affects those in the workplace while farmers worry about its impact on their crops or livestock. People living directly on a truck route for radioactive or chemical wastes may be concerned about spills or explosions, while those who have their own wells may fear contamination of their water supply.

After listing each sector of the population that might support the environmental organization and the two or three messages that are likely to motivate it, planners of the educational campaign decide how to reach each group. Some choices are obvious: parents can be reached through the schools and PTAs; workers through the unions; and farmers through agricultural associations. Finding groups that have already organized people with related concerns can greatly speed the process. Kansans for Safe Pest Control, for example, reached those farmers they thought would join their fight against pesticide spraying by contacting organizations of organic farmers and beekeepers. Opponents of a toxic waste facility to be built on an undeveloped rural site in Pennsylvania used a rod-and-gun club to reach hunting and fishing enthusiasts in the area. Both the Greater Newark Bay Coalition Against Toxic Waste and the Coalition for a United Elizabeth sent speakers to union meetings of those firefighters who would have to respond to an emergency at the toxic chemical waste facilities the groups opposed.

The fourth step is to develop educational materials. Some groups have found it useful to prepare a few pamphlets, fact sheets, or position papers for general distribution and then to develop materials aimed at one or another specific group. In order to decide how to reach a group most effectively, educators must assess how its members learn. Do materials need to be in more than one language? Some groups in New York and Los Angeles prepare their leaflets in English, Spanish, and Chinese. What reading level is appropriate? Should the material be in the style of newspaper articles, comic books, or photonovellas, or is a videotape more likely to persuade people? By knowing what formats people are familiar with, organizers can tap into existing channels of communication.

The final task is to plan a timetable for the campaign. An effective strategy is to first approach those groups most likely to support the environmental organization's objectives and then use this support to reach other sectors of the population. Trying to do too much too quickly will burn out educator-activists and can lead to sloppy work. On the other hand, too slow a pace can dissipate energy and enthusiasm, derailing the campaign before it begins. Some groups have used a seasonal approach to education. Stop Project ELF, the Wisconsin group opposing a navy communications system, used May and June—a peak visiting season for out-of-towners—to educate tourists whose resort areas might be damaged by the project. Similarly, the anti-Shoreham nuclear plant coalition hired a pilot to fly an airplane carrying a banner reading "Radiation Kills—Stop Shoreham" over eastern Long Island's beaches on August weekends.

Methods of Community Education

Effective educational campaigns require diverse methods of reaching people. The more educational activities a group sponsors, the more people it is likely to win over. In this section we consider five methods of community education.

Community Forums

Many groups organize open meetings to discuss a local environmental hazard. This is a good way to inform present or potential supporters about the issues, as well as to recruit new members. Several formats have been used. Antinuclear activists have sometimes sponsored debates with representatives of the nuclear power industry. By inviting sympathetic scientists or others with special expertise, the group can directly challenge corporate or government arguments. Many groups report that industry or government representatives are often so ill-informed or deceptive that they discredit themselves. An official of Dow Chemical, for example, testified at a California public hearing on DBCP (a pesticide that can cause male sterility) that despite animal research showing testicular damage from the pesticide, it had not occurred to Dow that it might also harm human male reproductive ability.[8] The controversy that a debate generates can also attract more people than does a one-sided presentation: people like to hear both sides of an argument and then make up their own minds.

Other groups have chosen a less confrontational approach. At a teach-in, for instance, scientists, physicians, government officials, politicians, and community leaders may be asked to present their views. A question-and-answer session lets citizens air their fears and obtain information. Inviting a prominent public figure will draw people who might not otherwise attend. The Coalition for a United Elizabeth, for example, asked the environmental scientist Barry Commoner to speak at a public meeting on the health effects of toxic chemicals. Some groups have conducted traveling teach-ins. The Greater Newark Bay Coalition led a tour of environmental hazards, stopping to educate residents living near a toxic chemical warehouse, the burned-out Chemical Control facility, a proposed toxic dump site, and a polluting factory.

Still another format is a speak-out. An open microphone offers people an opportunity to speak their minds while a facilitator focuses the discussion and summarizes points of agreement and disagreement. A speak-out can be a valuable way to demonstrate that an environmental hazard transcends any one individual's problem. If people think their illnesses are their own fault—because of excess smoking or drinking or too much stress—

hearing that others face the same problem can be a powerful antidote to blaming themselves. When several mothers of children with cancer spoke out at a public meeting in Rutherford, many parents were convinced that the threat they faced was a community problem, not only a personal one.

Some environmental groups have been able to force a government organization to sponsor a forum, lending credibility to the event and increasing the number of people reached. A Wisconsin antinuclear group, for example, convinced the county extension office to organize a public meeting on a proposed nuclear site. The testimony of state experts on the dangers of the facility contributed to public opposition to the project. On the other hand, an agency that adamantly denies that there is a problem may stack a meeting it sponsors in order to bolster its own position. At a Staten Island meeting set up by a community group and several city agencies to discuss chemical contamination of a Department of Sanitation landfill, more than a dozen city experts and politicians spoke before residents had a chance to ask questions. By then, those who had come to learn or to challenge the city's analysis of the problem had either fallen asleep or become so angry that rational discussion was impossible.

Public Speaking

Community forums reach people who are already sufficiently interested in an issue to come to a special event. Sending speakers to regular meetings of community organizations, on the other hand, gives environmental activists an opportunity to educate those who might not yet know about an environmental hazard. Wherever people gather, organizers can educate. We Who Care in Rutherford and the Harlem group that forced the asbestos cleanup used the school PTAs as the starting point for their educational campaigns. The Martinez Environmental Cooperative in California, which sought to end toxic discharges from oil refineries, sent speakers to a meeting of the union representing the workers in these plants. Foes of a chemical dump site in Pennsylvania spoke at a get-together of a trout fishermen's club whose stream was in danger of contamination. Folks Organized for Responsible Energy, a Minnesota group working to ban uranium mining, went to church groups to preach its message about the dangers of radiation.

Speakers at such meetings have several goals. First, they want to alert people to the danger posed by the hazard. Finding a way to portray the danger graphically or visually increases the likelihood of convincing the audience. A chart showing the number of trucks with hazardous cargoes crossing a busy intersection every day helped the Washington Heights group demonstrate the size of that hazard. Coastal Carolina Crossroads used copies of newspaper articles showing increased cancer death rates around oil refineries to document a potential hazard to the public.

Second, speakers at community meetings have greater success when they focus on the *specific* way a hazard affects the people at the meeting. Opponents of nuclear waste shipments can show parents' groups the proximity of a highway route to a school. Union members can be shown the danger a chemical may pose in the workplace. Farmers need information on the effects of toxic substances on their crops and livestock.

A third goal for speakers at meetings is to let people know what they can do about the hazard. This means educating them about the causes of the problem and suggesting political strategies for its solution. Good organizers never leave a meeting without having made several specific suggestions: read a pamphlet, sign a petition, talk to neighbors, come to a demonstration, join an environmental group. In addition, proposals should be made for what the community group as a whole can do: endorse an environmental group's demands, send a letter to public officials or a newspaper, or agree to donate resources such as volunteers, a mailing list, or funds. If it is presented with a number of options, the community organization can do whichever it feels comfortable with. A firm promise of limited support is more useful than a shaky vow of total commitment.

Community Events

Speaking to people is one way of educating them. Sometimes, however, action can communicate more clearly and forcefully than words. Direct action has several goals: to make people aware of a problem, to educate them about its causes, to put pressure on the authorities to remedy the situation. In this section, the discussion will focus on direct action as an educational method; chapter 8 will discuss direct action as a demonstration of political power.

In North Carolina hundreds lay across a road to block trucks from dumping PCB-contaminated soil at a county landfill. In Tennessee several community groups posted signs along the banks of a river warning people that chemical waste made the water unsafe for swimming and the fish dangerous to eat. An Atlanta group organized a caravan to follow the route of a nuclear weapons shipment, speaking and distributing leaflets on the dangers of nuclear materials transportation as they went. Another antinuclear group, this one in Wisconsin, put a huge banner across the town's main street proclaiming "Nuclear Power Poisons."

Actions such as these educate a community in several ways. The drama and excitement of the event get people talking about the issue, and a group's request for information and speakers often increase after an action. Actions also force the authorities to respond. The local utility company protested when the Wisconsin group displayed its banner, and the media covered the dispute. Everybody in town had to ask themselves, "Is it true that nuclear power poisons, or is the utility company correct in asserting that nuclear power is harmless?"

Action can also show people what is at stake and who is responsible for the problem. A Long Island antinuclear coalition demonstrated at a Nuclear Regulatory Commission hearing on the Shoreham reactor, bringing with them several truck loads of potatoes, cauliflower, and live Long Island ducks. The produce illustrated the danger the nuclear power plant posed to the area's agricultural economy. Several antinuclear groups have released helium balloons near nuclear power plants. Attached to each balloon is a card with the message "This balloon symbolizing radioactive fallout was released near (name of facility) on (date). Radiation, even in tiny amounts, has been shown to cause cancer. Please let us know how far this balloon traveled by mailing back this card."[9]

In Newark, New Jersey, opponents of a proposed toxic waste disposal facility demonstrated at a dinner given in honor of the board of the regional authority that was planning to build the plant. Through newspaper coverage of the demonstration, people learned about who had made this decision for their community. Environmental groups have also used street theater, placarded hazardous materials routes, dumped garbage on the front lawns of the headquarters of polluting companies, and staged sit-ins at the offices of corporate or government officials.

Activists who want to use action to educate need to choose a clear message, select a target that will embody that message, and then plan the event carefully so that it will have the desired effect. Since an action that fizzles can be discouraging, it is wise to choose direct action only if the resources and commitment to execute it well are readily available.

Written and Audiovisual Materials

Nearly every environmental group at some point produces and distributes flyers, fact sheets, or pamphlets. Such literature can make more complex and sophisticated arguments than can a public speech or demonstration. People can take a pamphlet home, read it again, or show it to a neighbor. It can reach people who might never attend a meeting.

What distinguishes a good piece of literature from a bad one? First, it has a clear message. Like a well-written newspaper story, a pamphlet's headline summarizes the main point and the lead paragraph then gives the reader the most important facts. Second, it is written in language its intended audience understands. For example, a New York City council member once suggested at a conference for environmentalists that all written materials prepared for elected officials should be in large type with many photos. Otherwise, the politician suggested, the material would get lost in the piles of literature that these officials routinely receive. Third, the layout should make the material easy to read. Photos, charts, drawings, or cartoons can help, while a solid page of single-spaced type can discourage even the most determined reader. Finally, every piece of literature should include sources of further information and suggestions for action. This might be the group's address and phone number, or the names and telephone numbers of officials to whom complaints should be directed. For example, West Virginia Citizens Against Toxic Sprays included on its leaflets the number to call if exposed to toxic spray, how to test a garden for aerial drift of pesticide, which government and industry officials to complain to, and how to join or contribute to the group.

Environmental groups have found many imaginative ways to present written information. A Berkeley, California, ecology group put "parking tickets" with a message about the automobile's contribution to air pollution on the windshields of cars parked on the street. Stop Project ELF in Wisconsin handed out a green bill-sized leaflet with "$1 million" printed on it. The leaflet went on to say: "This $1 million is just a tiny amount of the federal tax money wasted on Project ELF. $150 million has already been spent. Hundreds of millions more would be spent to construct an operational ELF system. This is your money. On the back print your name, address, phone, and a short message as to how you'd rather see this money spent. Return it to us and we'll send it to Washington, D.C." BLAST on Staten Island wrote and distributed a fable about an "imaginary" place called Apathy Island, which was flattened by an LNG explosion.

Groups preparing their own educational materials need not limit themselves to the printed word. Slide shows, videotapes, films, and posters are all useful methods of community education. Although these are more expensive to produce than flyers or pamphlets, their emotional impact makes them powerful educational tools. Each medium has its own particular advantages. The equipment for producing and showing slides (a camera and a slide projector) is widely available, while videotapes, which are shown on a television screen, are a familiar medium for many. If the video has been filmed in the community, viewers are excited about seeing their own surroundings and neighbors on television. Some environmental groups have collaborated with filmmakers or film students to produce audiovisual materials. The resources and skills available will determine which method is chosen.

Mass Media

"When we got on television, suddenly our activities gained great credibility," reported a member of a Pennsylvania group fighting toxic waste dumping. "Evidently that's when friends and relatives and neighbors decide you're not crazy." "We feel our frequent visibility in the press constantly causes the people of New Jersey to examine nuclear issues," explained an activist from the Safe Energy Alliance, which helped to win a seven-year ban on uranium mining and exploration.

Newspaper, radio, and television coverage is an essential tool for community education. It is one of the best ways to reach a significant portion of the population and can give a group visibility and credibility. Environmental groups have employed many methods to get their messages onto the airwaves or into print, ranging from letters to the editor, paid advertisements, and public service announcements to news coverage of demonstrations, interviews on talk shows, and press conferences. Every organization, no matter what its size, budget, or technical expertise, can find some way to get its message into the media.

Many groups find it beneficial to assign responsibility for this task to one person or to a small committee. This helps prevent contradictory information from going out or the loss of press contacts due to poor follow-up or missed opportunities to get coverage. Once a few people have developed good skills in media work, they can then train others. A log of media contacts and a scrapbook that includes each of the group's appearances in print or on the airwaves helps keep track of progress in getting the message across. The log can also help evaluate the relative success of, say, news articles, talk show appearances, or radio public service announcements in generating new members, invitations for speakers, or requests for information.

Getting media coverage for the first time is probably the most difficult task. Until the group's name and the issue it is fighting make the news it is not "certified" as a newsworthy organization. Local environmental groups have used a variety of techniques to break out of this Catch-22 situation. BLAST on Staten Island, for instance, got its first citywide publicity in an article published in the New York *Daily News*. The article was written by a journalism student whose professor was so impressed by her story that he urged her to submit it for publication.[10] By taking the time to talk to the student, BLAST activists had won important coverage. Other groups have found it easier to get articles into the local newspapers than the larger urban ones; sometimes a few stories in the local paper may lead larger circulation publications to pick up the issue.

Timing may also determine whether a story is picked up or dropped. The planners of a successful reproductive-rights media campaign in the Bay Area recommended that groups tie their campaigns to events that will improve their exposure and effectiveness.[11] The Love Canal Homeowners' Association, for example, used the extensive news coverage that candidates for governor and then president attracted to get their message to statewide and national audiences. In one appearance on a national television show during President Carter's reelection campaign, Lois Gibbs urged viewers to send telegrams to Carter demanding federal funds for resettlement. The White House received 3,000 telegrams in twenty-four hours.[12]

Reporters are more likely to pick up a story that has already gotten some media attention elsewhere than to cover a new problem. Unfortunately, a disaster that has already occurred always makes better news than a potential danger. The leader of the Harlem asbestos fight believed that TV news coverage of similar school asbestos problems in nearby New Jersey made reporters more willing to see the Harlem school's plight as worth covering.

Still another tactic to win coverage is for a local group to join forces with a large national organization. A press conference or news release co-sponsored with the Sierra Club or the Audubon Society is more likely to attract attention than one initiated by an unknown organization. The local group can then use the contacts and credibility it has achieved to get coverage on its own. Inviting a prominent expert or activist to speak at an event will also increase the likelihood of making the news.

Once an organization hits the headlines, its next task is to develop strategies for winning ongoing (and favorable) coverage. Holding firmly to several principles will help achieve this. First, total honesty with journalists is a prerequisite for credibility. When reporters learn that an environmental group and its expert consultants can be trusted to give accurate information, they will regularly turn to them for reaction to environmental problems. Professor David Wilson, a Vanderbilt University chemist who frequently advises community groups on toxic chemical issues, has observed that "if you're sloppy or overdraw conclusions, you lose credibility with journalists. And credibility is all you have."[13]

Second, a group must learn what reporters consider newsworthy. The media coordinators of the Clamshell Alliance observed that "it is vital not to pester the press incessantly with non-news or rhetorical overlong statements or become consumed as an organization with media attention."[14] The average newspaper gets over two hundred press releases every day; unless its editors see something that grabs their attention, most end up in the wastebasket. This means that sometimes the reporters themselves must be a target for education. By giving them background information and access to expert consultants, and by demonstrating the impact of an environmental hazard on ordinary people, it is possible over time to win more sympathetic coverage. A leader of We Who Care explained how her group's ongoing efforts changed the kind of environmental stories that appeared in local papers. "Our fight opened up a whole new avenue of reporting that had not been popular. Editors at the time, we were told by reporters, would not allow environmental problems to be published. Now not a week goes by without something about hazardous waste sites. The papers really did a kind of turn-around. It's probably related to what we continue to do and to the fact that it seems to generate a lot of interest in readers."[15]

Another method of getting regular coverage is to give reporters new information or new angles. The Love Canal Homeowners' Association organized what they called the "horror story of the day." Each day another family volunteered to describe to journalists the illnesses they believed could be attributed to Hooker Chemical's abandoned wastes. The feature stories based on these interviews helped to convince millions of readers around the country that the Love Canal residents deserved compensation for their losses.[16]

Accurate news tips can also help to keep a story on the front page. Opponents of the Shoreham nuclear power plant on Long Island tipped off reporters to the discovery of serious safety problems, leading to continued coverage. And journalists have often helped to expose illegal toxic dumping. Careful preparation for meetings with media representatives ensures that a group's story will get through. The Clamshell Alliance media committee practiced answering hostile or skeptical questions before appearing at press conferences.

For groups that cannot get regular news coverage, there are other ways to use the mass media. Letters to the editor, guest editorials (or "op ed" pieces), public service announcements, cable television programs, and paid advertising can all be used to get a message to the public. The anti-Shoreham coalition, for example, bought ads in a local newspaper using the same piece of art—children playing and farmers working—as the utility used in its pro-nuclear advertisements.[17] The text of the message rebutted the power company's contention that its evacuation plan was realistic. A Montana antipesticide group used local public-access cable television to give its members a forum in living rooms throughout the region.

Useful as the media have been in educating about environmental hazards, depending on them to communicate with the public can be dangerous. Coverage may not be fair: Citizens Against Toxic Sprays in Olympia, Washington, reported that "the local papers are hostile. They print almost nothing we submit. If they do, it is cut, rephrased, misquoted, and garbled." The group attributed this hostility to the timber industry's influence on the media. Similarly, the Rusk County Citizen Action Group in Wisconsin claimed that local coverage of their actions was "skimpy and slanted." An antinuclear group in New York charged that the media "avoid and whitewash the issue." Clamshell Alliance activists reported that "despite major efforts to focus media attention on the issues (nuclear power and safe alternative energy), reporters repeatedly concerned themselves primarily with the surface aspects (number of participants, how they were dressed, etc.) and often only came close to addressing the issue by reporting the slogans on picket signs carried in a march."[18] Television reporters, who depend on visual images, have a particularly hard time addressing substantive environmental issues.

Deadlines, market pressures, and narrow ideas as to what constitutes news also encourage superficial coverage. In order to submit stories that are acceptable to their editors, journalists look for a "frame" for a story, then report all subsequent events from that viewpoint.[19] Themes or frames that often appear in environmental stories include innocent victims (usually women and children) of toxic chemicals; ordinary citizens expressing fears not substantiated by facts ("environmental hysteria"); corrupt or ignorant industry or government officials whose individual misdeeds have caused problems; and concerned officials doing all they can to remedy the situation.

While these media clichés might seem to work in favor of environmentalists, in fact they present problems. Having heard the same story many times before, the audience simply tunes out and the repeated assurances that officials are doing everything possible convinces people that they can sit back while the "experts" take care of the problem. Helene Brathwaite, the leader of the Harlem school abestos struggle, explained: "For some people our fight to get rid of the asbestos was a media event. If it weren't for the intense media coverage, many parents would have continued to work and work hard. But when the politicians and the superintendant were questioned on TV, they'd say, 'We're doing that.' 'We're fixing the problem.' So people felt there wasn't a need to continue to do something. In retrospect, I think the media coverage was necessary but it did make our work harder."[20] Most important, superficial coverage does not look for the root causes of this country's environmental hazards and thus does not help people to develop strategies to combat them.

The reasons for these problems are fairly obvious. Newspapers and television stations depend on industry for advertising, and are reluctant to challenge their corporate sponsors. Moreover, publishers and station owners usually come from the same socioeconomic class as corporate directors and managers, making them unwilling to criticize their friends and neighbors. Mass media outlets are increasingly being bought up by multinational companies, often the ones responsible for toxic pollution. For example, General Tire and Rubber Company, Gulf + Western, and Westinghouse Electric are among the fifty companies that own and control most major U.S. media.[21] Moreover, most big city newspapers share corporate directors with petrochemical and other polluting companies. The *New York Times* board of directors includes people who also sit on the boards of Sun Oil, Johns Manville, and Bethlehem Steel, while the *Washington Post* has on its board of directors people who also sit on the boards of Allied Chemical, IBM, and the Ford Motor Company.[22] Even in those situations where there is no such conflict of interest, editors and reporters must worry about the growing spate of corporate libel lawsuits against the media. Mobil Corporation, for example, recently purchased "defamation insurance" to cover the legal fees of its executives who press libel suits. Floyd Abrams, a leading First Amendment lawyer, noted, "What Mobil is trying to do is to so increase the potential cost of reporting about it that reporters and editors will conclude that it is not worth the effort."[23] Those who own and control the U.S. mass media put profit and corporate rights ahead of health concerns and democracy, so it is hardly surprising that they do not want to consider community struggles as news that is fit to print. Nevertheless, an awareness of the interrelationships between mass media ownership and the stories that are covered need not discourage activists from using the media: as the case histories illustrate, many groups benefited from good coverage. By understanding media links to big business, organizers can set realistic goals as to what to expect from this method of community education. Finally, activists need to remember that such education can never replace the more humble leaflets, discussions, and picket lines. A newspaper article or a TV show can make a person more receptive to new ideas, but it is the intensity of face-to-face interaction with an organizer, a concerned neighbor, or a fellow worker that moves a person to change her or his beliefs or to take action.

Obstacles to Community Education

One of the most difficult tasks facing the activist-educator is to help people to think for themselves. An organizer from Oregon summarized this problem clearly: "People are generally trained from childhood to be intimidated by those who hold higher status or authority and thus fail to use their own critical consciousness and simple powers of awareness, observation and initiation." This is one reason why it took months of educational work before the Love Canal Homeowners' Association could convince its neighbors not to trust government and corporate assurances that the toxic waste posed no threat to their health.

Several strategies can help people to begin to think for themselves. One is to expose clearly and rationally the misstatements or distortions spread by industry or government. The antinuclear movement, for example, has been remarkably successful in challenging industry contentions that nuclear power is the safest and cheapest source of energy. By documenting the thousands of accidents and construction errors, by contesting the industry's economic arguments, and by promoting solar power and other energy alternatives, antinuclear power activists have won broad public support for their goals. Another strategy is to show respect for people's beliefs and values, but to offer an alternative interpretation of the facts within their belief system. Many Americans firmly believe that they can trust statements by those in power and that important decisions are indeed made democratically. Those whose experience has led them to think differently are often frustrated by this apparent naiveté and try to bludgeon potential supporters into sophistication with example after example of abuses of power. A more effective approach is for the activist to say, in effect, "You are right to want to trust government leaders and you are right to want decisions about your community to be made democratically. Now, let us see if that's the way planning for this toxic waste facility (for example) has proceeded." By emphasizing those values and commitments that the people involved can agree with, educators open the door for real learning.

Still another strategy is to force people to compare their personal and collective experience with official statements. After the propane gas truck leak in Washington Heights, officials asserted that evacuation had proceeded smoothly and that the city was therefore prepared for such accidents. Those who lived in the area, however, knew that the emergency response plan had no provisions for warning the many residents who did not speak English, for helping handicapped and frail elderly people, and, most shockingly, had no plan as to where the evacuated people should go once they left their homes. At a public forum sponsored by the community coalition, the revelation of these discrepancies helped the group to convince people of the need for action.

Activists can also help citizens to develop critical thought processes. The long-range goal of educators is to help learners develop their own powers of analysis. Environmental activists want not only to convey facts about this or that hazard but also to empower people to defend their communities against future threats. By teaching people how to investigate a problem, how to analyze its causes, and how to plan a strategy for change, organizers can ensure that their efforts will have a lasting impact. To accomplish this goal, a group must establish an ongoing dialogue with potential supporters that will allow them to take one intellectual step at a time. Changing people's consciousness, their priorities for spending their time, or their political perspective can be a slow process so it is unrealistic to expect that long-held attitudes will change overnight. But creating opportunities for open discussion, by respecting those who disagree, by using a variety of educational methods and arguments, activists have found, over time, that they can win public support for their goals.

Every community educational campaign at some point encounters apathy, denial, or hostility. An apathetic response from the public wears activists down, leading them to ask why they are making the effort. Hostility isolates organizers, who may even become antagonistic to potential supporters. Denial—the refusal to acknowledge that a problem exists—can make activists feel that

TABLE 3 Obstacles to Community Education: Apathy

Cause	People Say	Activists Can
Too many other commitments	"I believe you that the problem is bad, but I'm so busy with school, job, family, etc., that I can't be bothered."	1. Help to reduce other commitments (e.g., provide child care for meetings) 2. Scare people with consequences of inaction
Previous failures in community action	"I agree that it is dangerous, but I've gotten involved so many times and it's never made a bit of difference. You just can't fight city hall."	1. Invite those who have had victories to describe their success and help plan strategy 2. Help to analyze specific causes of past failures
Lack of information documenting negative impact of hazard	"Look, I'm sure you're a very nice person, but I'm just not convinced that it is dangerous. Why waste my time on a minor problem?"	1. Portray hazard dramatically, e.g., invite victims to testify 2. Invite scientific experts to document magnitude of hazard
So many prior threats that new dangers are not believed	"Every day I hear about something else that causes cancer. If I got upset about everything, I'd spend my whole life fighting. After all, we're all going to die anyway."	Invite those who have succeeded in eliminating a hazard to share their success

they are crazy, living in a world where only *they* perceive the real dangers.

To surmount these obstacles, their causes must be accurately diagnosed. Tables 1, 2, and 3 describe some of the reasons for apathy, denial, or hostility and suggest some strategies for countering them. Lois Gibbs, for example, told parents who claimed they were too busy to get involved that failing to take action to protect their children from exposure to toxic chemicals was just as much child abuse as beating them up.[24] But it must be remembered that for fear to be an effective antidote to apathy, the threats must be real and a clear course of action to remedy the hazard must be offered.

Apathy is an understandable response to the experience of powerlessness. If every previous effort at change has led to failure, why continue to beat your head against a wall? Coastal Carolina Crossroads, the group fighting a proposed oil refinery, put out a pamphlet describing how eleven other communities around the nation had successfully blocked refineries. Stories of triumphs, demonstrations that you can fight city hall and win, help to overcome the "don't have no power" blues.

Psychiatrists describe denial as a mental process by which sources of anxiety are ignored. It can be an appropriate response to threats that cannot be eliminated—the inevitability of one's own death, for instance; but it is counterproductive, even suicidal, when used to deny a source of anxiety that *can* be eliminated. Yet for some people, to accept that an environmental hazard threatens them with serious illness or death challenges some of their most cherished beliefs: the belief in their ability to protect their children and families from harm, the worth of the years of work that went into buying their homes, the government's ability to protect them, and their capacity to control their own lives. It is hardly surprising that it is sometimes easier to deny that a problem exists than to reevaluate one's view of the world and the way it works. The reactions of some residents of Oak Ridge, Tennessee, to the news that Union Carbide, which operates the federal government's research laboratory and manufacturing facilities there, may have polluted a creek with mercury illustrate the process of denial. E. L. Henderson, who works at the Oak Ridge plant and swam and fished in the creek as a child, said he was not concerned about the reports on contamination. "I'm not

TABLE 4 Obstacles to Community Education: Denial

Cause	People Say	But Think	Activists Can
Fear that damage is already done	"It couldn't get me, I've never been sick a day in my life."	"Maybe I have cancer already, I'd rather not think about it."	1. Stress possibilities of prevention 2. Help people to get accurate information on current health status
Fear that problem is insoluble	"Oh, that could never happen here."	"If (for example) my water is poisoned, there's nothing I can do about it."	1. Help people understand specific causes of problem and plan appropriate interventions 2. Present success stories
Fear that admitting hazard would disrupt world view	"I can protect my family."	"If I haven't protected my family, what good am I?"	1. Help people understand real protection of family requires group, not individual, action
	"I can trust my government."	"If I can't trust the government, who can I trust?"	2. Help people to build organizations they can trust

worried about the mercury," he observed, "but I would worry about it less if they stopped writing about it."[25]

An important first step in breaking through denial is to identify and acknowledge legitimate fears. Public speak-outs can help to bring the anxiety that feeds denial to the surface: once it is clear that many people share certain fears, neighbors and friends can develop emotional support systems to help each other. A second strategy for combating denial is to make the hazard more concrete, more imaginable. Physicians for Social Responsibility, an anti-nuclear organization, has helped people to confront the horrors of nuclear war by describing in great detail the effects of a nuclear explosion on a particular city. Photos and testimony from victims of the Hiroshima and Nagasaki explosions have been used to force people to visualize the result of nuclear war.

Finally, and perhaps most important, helping people to understand that they can do something to change what they fear can help to cut through denial, as it does with apathy. By reminding residents that communities across the country have stopped the construction of nuclear power plants, forced the cleanup of toxic dump sites, or banned aerial spraying of herbicides, organizers can convert denial into anger and then commitment.

Every group that enters an environmental fray encounters some hostility. As shown in Table 3, the reasons can range from feeling one's financial status is threatened to fear of losing a job to anti-communism. As with apathy and denial, a correct diagnosis of the causes of hostility makes finding the appropriate response easier. "We reacted to dissent by talking logically and giving people the right to disagree. We tried not to be arrogant and butt heads," reported a member of the Price County (Wisconsin) Nuclear Education Association, a group fighting to block a nuclear waste dump. Respectful dialogue and debate is probably the best way to meet disagreement. It leaves the door open for further discussion and does not polarize the situation prematurely.

Such a patient response is often rewarded later. A group fighting water pollution in Minnesota reported that the local government was at first "antagonistic" but that with time it became "very responsive" to the group's efforts. Similarly, as noted earlier, We Who Care found a significant improvement in media coverage of environmental stories over time.

It is probably not productive to engage in debate those who have

TABLE 5   Obstacles to Community Education: Hostility

Cause	People Say	Activists Can
Feels job threatened	"You kooks are going to cost me my job. You care more about wildlife than working people."	1. Describe how hazard affects working people 2. Include demands for job protection (see chapter 10) 3. Explain "environmental blackmail"
Has stake in status quo	"It costs too much to clean up my factory. You people are antiprogress."	1. Ignore 2. Respond with political strategies (see chapters 7 and 8) 3. Present information that will show economic benefits of change
Opposes any effort at change	"You're all a bunch of commies. Why don't you go back to Russia."	1. Ignore 2. Explain to others how red-baiting is used to oppose any changes
Thinks your tactics/strategies are wrong	"I agree with your goals but I think you're doing more harm than good."	1. Ask for and consider alternative suggestions 2. Suggest he or she act on his or her beliefs in a way that seems appropriate
Doesn't want another worry	"I have enough to worry about. Leave me alone."	1. Leave person alone 2. Suggest that hazard could make other worries irrelevant

already firmly decided that an environmental group is wrong. A leader of Coastal Carolina Crossroads explained that "we did not try to convince individuals adamantly opposed to our views. We spent our time and efforts supporting those on our side and presenting the factual materials we had to those who had not taken a position against our views."

Not all hostility stems from those who misunderstand or disagree with activist aims, however. The most potent opposition generally comes from those who have a stake in the existing situation—the corporate owners or managers whose profits would be lowered by spending more on pollution control, the politicians who depend on corporate campaign contributions, and the government officials whose agencies have to pay for the cleanup. For these individuals and institutions, an environmental group's response to hostility cannot be educational but must be political. The task is not to change these people's minds, but to win enough power to force them to act differently.

Community Education: A Means and an End

Every stage in a community's environmental struggle is both a means to an end and an end in itself. Education helps mobilize people to act to eliminate a hazard, and it also makes people more knowledgeable, more skillful, and more powerful.

Many local groups, therefore, considered their educational campaigns significant accomplishments in their own right. "We've definitely heightened public awareness about pesticide spraying," states a member of the Chippewa Valley (Wisconsin) Friends of the Earth. "We have alerted others to Pennsylvania's lack of compassionate, reasonable siting criteria for toxic waste facilities and brought to light that such facilities are purely for personal gain and are not really needed," said another activist. The leader of an antiherbicide group in Oregon acknowledged that as a result of their efforts "the tremendous power and authority of the governmental-industrial complex has not been diminished," but that "at least we have all become more personally aware of the extent of that power."

Even if environmental groups fail to eliminate the hazard they are fighting, their success in raising people's consciousness contributes knowledge and skills that will be an asset in the next battle. People who understand more about the dangers they face and about who is responsible are better equipped to protect themselves.

For every organization, the educational campaign is a critical stage in the struggle for a healthful community. The support that activist-educators garner through community education provides the foundation for their political power. In the next two chapters, we explore how groups can use that power to win their goals.

7. STRATEGIES FOR ACTION:
Legal and Legislative

Case History 1: Southern Utah

Every winter since the 1870s ranchers in southern Utah have been driving their sheep into the Nevada desert.[1] In the early 1950s the U.S. government began to test nuclear weapons at the Nevada Nuclear Test Site, only forty miles from the sheep's winter quarters. In small towns in Utah, families would rise at dawn to watch the detonations. McRae Bulloch, a rancher from Cedar City, Utah, took his sheep into Nevada in November 1953. "We'd be sitting there in the sheep wagon before dawn," he explained to a reporter recently, "and the whole sky would light up. We'd say, 'They've set off another bomb.' It came up in smoke and dust and mushroomed out, all kinds of colors, and drifted whichever way the wind would carry it. You couldn't feel the radioactive dust, but the air looked cloudy, hazy."[2] From then until 1962, more than one hundred atomic bombs were exploded at the Nevada site.

In the spring of 1953, Utah ranchers lost more than 4,500 sheep to a strange disease characterized by sores on the animals' mouths and ears and the birth of underweight lambs, some without any wool. These losses bankrupted many ranchers. In 1956, Federal Judge A. Sherman Christensen ruled against the ranchers' lawsuit, which charged that their sheep's deaths were due to exposure to radiation. Twenty-six years later, in response to a new suit by the ranchers, the judge reversed his decision. Citing reports of deliberately concealed evidence, pressured witnesses, and deceitful conduct, he declared that the government had "perpetrated a fraud upon the court" and he went on to charge that "the processes of the court were manipulated to the improper and unacceptable advantage of the defendants."[3] (Despite these findings, in 1983 an appeals court reinstated the original decision against the ranchers.[4])

159

At about the same time that Judge Christensen reversed his decision, another case came to trial in Federal District Court in Salt Lake City. In that suit 1,192 present and former residents of southern Utah charged the federal government with negligence in carrying out its above-ground atomic testing. They demanded millions of dollars in damages for the deaths and illnesses that they argued were due to radioactive fallout from the tests. Their suit listed three hundred cancer cases of whom lawyers singled out twenty-four for consideration in the trial. Nineteen of the victims had already died, while the remaining five were undergoing treatment. Elmer Pickett, a storekeeper from St. George, Utah, had lost his wife, a sister, a five-year-old niece, and eight other relatives to cancer since 1954. He described the bitterness many residents felt: "Our own country did to us what the Russians couldn't. We were just classified as expendable while they done their experiments and played with their toys down there. We have been wrongly used by our government. They lied to us about the dangers."[5]

In September 1982 the trial began. Forming the Committee of Survivors, the plaintiffs and their lawyers used the court proceedings as a forum to expose the government's negligence. Dr. Joseph Lyon, a cancer researcher from the University of Utah, testified about a study he had published in the prestigious New England Journal of Medicine that showed that the rate of leukemia in southern Utah during the most intense period of testing was 2.4 times higher than it had been before and than it was after. The rate in the Utah counties closest to the test site was even higher. Dr. Lyon concluded that the "most likely cause" of the excess leukemia deaths was radiation, which previous studies had shown to cause leukemia in two to ten years.[6] Another witness, a man who had monitored the amount of radiation near the bomb site, testified that the instruments used to measure radioactivity "went off scale" during some of the tests.[7] He also reported that one of his jobs had been to set up roadblocks where cars traveling through the area after a test were decontaminated. In courtroom testimony, Dr. Karl Z. Morgan, former director of health physics at the Oak Ridge National Laboratory, charged that the government's safety measures were inadequate and "not in the spirit" of what was then known about radioactivity's perils.[8]

The government challenged the Committee of Survivors' claims on several fronts. Assistant U.S. Attorney Ralph Johnson, for instance, suggested that the excess leukemia rates might have been due to stress: "Stress is one of the factors that simply must be considered," he argued, even though there is no scientific evidence linking stress with leukemia.[9] Another U.S. attorney called the plaintiffs' testimony "ludicrous" and argued that since injury claims had not been filed within two years of the date at which the damage was done, as the law requires, the suit was not valid.[10] Other lawyers argued that the government was immune from lawsuits where matters of national security were concerned.[11]

In May 1984 a federal district judge ruled that the fallout from the nuclear tests in Nevada had caused nine people to die of cancer and that the government was guilty of negligence in that it had conducted the tests. Although the judge's decision was less sweeping than the cancer victims had hoped and although the government appealed the decision, nevertheless the case set an important precedent that will benefit uranium miners, veterans, and other groups seeking compensation for illnesses resulting from exposure to low-level radiation.[12]

Case History 2: Elizabeth, New Jersey

A case history in chapter 1 described the battles between the Coalition for a United Elizabeth (CUE) in Northern New Jersey and the chemical and toxic waste industries that were polluting that town.[13] In an interview, Sister Jacinta Fernandes, an organizer of CUE, explained why her group decided not to file a lawsuit against Chemical Control Company, the owners of the warehouses that exploded in April 1980.

> After the fire at Chemical Control we were meeting with the Rutgers University Law School Constitutional Clinic. They had a whole team of students that worked with a lawyer, and for a whole year they researched possibilities of suing Chemical Control. But the more they got into it, the more gigantic the thing seemed, because they found that with all the other industry in the area, they could not sue just Chemical Control without involving all those other industries, all of which have very high-powered lawyers to represent

them. Also, we brought in doctors and epidemiologists who said that to do testing of the residents was very complicated because they would have to develop a specific test to find out every chemical they were dealing with. Since there are so many known and unknown chemicals, they'd have to be monitoring people over a long period of time, which would require a terrific amount of funding.

So after a whole year of meetings with doctors, epidemiologists, scientists and lawyers, and having the lawyers and students research where to get funding for such a lawsuit, the residents, the lawyers and the staff at CUE together came to a decision that that kind of lawsuit was not feasible. CUE staff just didn't have the time and energy for that type of effort. The lawyers told us we would have to be willing to devote full time to this. And the residents felt that they just didn't have the stamina to continue with that kind of thing. They were anxious to get some health testing, to get some results now.

Our decision not to pursue the lawsuit was also a political decision. We had other issues we were working on and even the toxic waste issue involved other things we wanted to move into. There are also other legal cases against Chemical Control that we can get into as a friend of the court, which will be easier for us to handle.

Case History 3: Mendocino County, California

On June 5, 1979, voters in Mendocino County in northern California voted 8,644 to 4,980 in favor of banning aerial spraying of phenoxy herbicides such as 2,4,5-T, 2,4-D and other substances containing the deadly chemical dioxin.[14] The referendum marked the first time that California voters had used the ballot to decide the question of herbicide use.

Since 1973, when more than four hundred people petitioned the county government to ban spraying on county roads, the use of phenoxy herbicides had generated controversy in Mendocino County. A few years later apple growers complained that their previously perfect apples were now growing "deformed, kind of crooked and gnarled," a condition they attributed to spray drift from the herbicides sprayed on nearby commercial forests to kill or stunt unwanted hardwood trees. In 1977 several Mendocino residents complained of flulike symptoms immediately after the timber company's planes sprayed near their homes. A year later a group of doctors at the county hospital issued a statement stating

that they "view with alarm the continued aerial spraying of
phenoxy herbicides. Scientific study has documented the potential dangers over the long term use of dioxin."[15]

Supporters of the spraying rejected these contentions. The
president of the Mendocino County Farmer's Bureau charged that
the antispraying advocates were "people who moved to the
mountains to do away with anything that has to do with civilization. They are trying to instill their philosophy on the county."[16]
Others claimed that the referendum was unnecessary because of
the Environmental Protection Agency's (EPA) recent temporary
ban on the use of two major phenoxy herbicides. Furthermore,
California Attorney General George Deukmejian ruled that the
referendum was illegal since federal and state laws already regulated herbicide use, and a vote was thus a waste of the taxpayers'
money.

Despite these criticisms, the referendum was held and it won by
a wide margin. The local district attorney, who supported the
initiative, said the victory showed that voters had "lost faith in
the government's willingness and determination to protect
them."[17] Luke Breit, the local coordinator of the Citizens Against
Aerial Spraying of Phenoxy Herbicides, thought that the vote "rejected the notion that we can leave this in the hands of experts.
There is a basic distrust of agencies that are not connected with
local people, men in offices who don't have to drink the water. I'm
hoping that the trend is that we will take the power of our lives
back into our hands."[18]

Local officials, even former supporters of the spraying, were
influenced by the election results. Despite the fact that the referendum did not have the force of law, County Agricultural Commissioner Ted Eriksen, Jr., said he would deny all applications for
aerial spraying of phenoxy herbicides unless the courts forced
him to do otherwise. "If this is what the people want to do, right or
wrong, I don't see that we have any choice but to interpret that as
their wishes," he stated.[19]

Case History 4: Lincoln County, Oregon

In rural Lincoln County on Oregon's Pacific coast, the results
were different.[20] Here, in November 1980, two ballot measures

that would have restricted spraying of herbicides were defeated by a five-to-three margin. One would have banned the application of herbicides by air within five hundred yards of any body of water and the other would have banned any herbicide spraying within one hundred yards of a body of water. The ballot initiative was sponsored by the Lincoln County Medical Society, which was pushed into action by two physicians who had become alarmed at what seemed to be high rates of birth defects.

Why did the bans lose in Lincoln while a more stringent one succeeded in Mendocino County? Perhaps the most significant difference between the two was the extent to which industry organized against the measures in Lincoln County. A lawyer who lived outside the area organized a group called Lincoln County Citizens for Common Sense. The organization mailed a letter to county voters asking, "How would you like to be cited, fined $500 and have a criminal record just because you put weed/feed on your lawn?" Hiring a high-priced Portland public relations firm, the group placed articles and advertisements in local newspapers and on the radio and barraged voters with mailings warning them that the ballot measures would disrupt the local economy.

Soon another group, Doctors for Facts, entered the fray. The group began a direct-mail campaign to assure residents that phenoxy herbicides were safe. One pamphlet asserted that "no chemical is hazardous if we consume it or are exposed to it in small enough amounts." Although the letterhead had a Lincoln County address, not a single medical person listed in the group practiced or resided in the county.

A week before the election the lawyer leading the Citizens for Common Sense threatened to sue supporters of the initiative, alleging that the ads against herbicide sponsored by local physicians were false. No legal action was ever taken, but the threat damaged the local doctors' credibility.

How was this elaborate campaign financed? A required financial statement filed with the county clerk showed that Citizens for Common Sense had raised over $27,000, more than $20,000 of which had been contributed by the major timber companies with holdings in the area—multinationals such as the Georgia-Pacific Corporation, Weyerhauser, International Paper Corporation, and Champion International Corporation. On the other hand, Citizens for a Healthy County, the measure's supporters, had raised $1,452 from a benefit concert and private contributions. By raising almost twenty dollars for every dollar raised by opponents, by exaggerating the consequences of the initiative, by playing on people's fears about individual freedom and the local economy, and—most of all—by being able to expose people constantly to their message, Citizens for Common Sense and their corporate sponsors had been able to win a victory in Lincoln County.

Environmental activists differed in their assessment of the defeat. "I made the fatal mistake of being optimistic," said one supporter of the measure. "I counted on the voters' thinking. I thought they would trust their own doctors enough to read the measures, or at least to question the sources of what they heard on the other side. What mattered, what won their votes, was not what they heard but how many times they heard it."[21] Another activist looked at the defeat more philosophically.

> As more distressing effects of herbicide exposure accumulate in the public health record, industry's overkill tactics may backfire. Thanks to the huge scale of this campaign, voters in the future will approach the issue with far greater awareness of the controversy than they have in the past. An informed, aware electorate will be less susceptible to the misinformation and scare tactics that can only be effective on an ignorant populace. By using only temporarily credible tactics, industry traded a tactical victory for a strategic loss.[22]

As the case histories illustrate, legal and electoral strategies have led to both victories and defeats. How can activists use these strategies most effectively? How can they decide which method is appropriate in a given situation? This chapter begins with a description of each strategy, and then compares the common advantages and disadvantages of the two. Finally, it offers some guidelines for their use.

Legal Strategies

During the 1960s and 1970s environmentalists fought for and won new legislation to control pollution. One of the important rights provided by such laws as the Clean Air Act Amendments of

1970 and the Clean Water Act of 1977 was the right of citizens to initiate lawsuits against government agencies and corporations that were violating these acts. As a result, in communities across the country the response to an environmental hazard has increasingly been to sue the offender. These legal efforts fall into two categories. Some groups have used *injunctive actions* to prevent a government agency from, for example, issuing permits for a nuclear power plant or a waste water discharge plant or to prevent a corporation from taking an action that will damage the environment in violation of a permit. In some states, citizens also have the right to participate in quasilegalistic administrative hearings, giving them another way to block dangerous projects. Other groups used *damages suits*, in which the injured parties have sought restitution for economic harm done by the government or by a private entity. Such actions usually depend on a common-law interpretation of nuisance, trespass, and negligence.

These efforts have won a variety of victories. Most concretely, such suits have stopped a number of environmental hazards. For instance, in a 1970 landmark decision the Wisconsin Department of Natural Resources banned the use of the pesticide DDT in that state. The action followed a series of legal and administrative battles with environmental and citizens groups on one side and pesticide manufacturers—such as Allied Chemical, Olin Mathiesen, and Geigy Chemical and the U.S. Department of Agriculture—on the other.[23] This and other court decisions provided the impetus for the Environmental Protection Agency's 1972 ban on the use of DDT inside the United States. Lawsuits have led to blocking nuclear power plants, halting pesticide spraying, reducing air pollution, and closing many toxic waste dumps.

Just as important as the specific victories, however, are the changes in public consciousness that have resulted from these courtroom confrontations. The Utah trial, for example, contributed to a national debate on atomic testing. Millions of Americans learned about the military's testing program and the callous disregard for the health and safety of residents and soldiers. At a time of national debate on military priorities and nuclear weaponry, the trial and its publicity helped to discredit the idea that nuclear testing and government secrecy benefit the people of the United States; it also exposed the government as an unreliable protector of its citizens.

Victor Yannacone, the lawyer who led the legal challenges to DDT, explained the educational value of a trial: "Only in a courtroom can a scientist present his evidence, free from harrassment by politicians. And only in a courtroom can bureaucratic hogwash be tested in the crucible of cross-examination."[24] Yannacone used this crucible to ridicule the incompetence of federal and corporate officials who were ill-equipped to refute the evidence presented by the scientists who testified as to DDT's dangers. Media coverage of the proceedings familiarized a national audience with the issues.

Some trials have brought previously unknown hazards to light. As a result of a $15 million lawsuit filed by a landowner near the Rocky Flats nuclear weapons plant in Colorado (charging that the plant had polluted his property with plutonium and uranium), legal investigators discovered high levels of radiation in two elementary schools within twelve miles of the plant. A fire at the Rocky Flats plant in 1957 had sent a cloud of radioactive material across the Denver area but at the time a plant spokesman had told reporters that there was "no spread of radioactive contamination of any consequence," a contention that was disproved when it was found that the two schools were badly contaminated.[25] Groups fighting toxic waste dumping also use court action to learn about what substances a company dumps and where.

For other environmental groups, lawsuits have won dollars. New Mexico Citizens for Clean Air and Water reported that they had "collected damages from two companies to be used for future environmental work and to pay lawyers' salaries." In an out-of-court settlement Consolidated Edison of New York and four other power companies not only promised not to build a new power plant on the Hudson River, but also agreed to provide $12 million to set up a new foundation to sponsor environmental and public health research on the river.[26]

Even when lawsuits fail, they can still influence the outcome of an environmental dispute. An EPA study of the impact of citizen opposition to toxic waste facilities concluded that while "most lawsuits initiated by opponents to date have been unsuccessful in the courts," they have been "moderately successful as a delaying tactic. They have also added substantially to the sponsor's cost."[27] Similarly, a member of Middle Paxton Concerned Citizens, a Pennsylvania dump group, reported that fear of "legal entanglements for many more years" eventually forced a developer to abandon plans for a new dump site. Using lawsuits as a delaying device is particularly effective if public opposition is expected to grow. Early legal interventions by antinuclear groups helped to postpone construction of certain nuclear power plants until economic factors and political organization led to a halt in construction. Such delays can also give a group more time to mobilize on other fronts.

Finally, legal action, successful or not, can help to increase government and corporate accountability. "Our lawsuits have kept the government and industry on their toes," noted an organizer of a clean air group in Pittsburgh. And in November 1982 three environmental groups filed civil lawsuits against nine New York State companies, accusing them of "systematically and significantly" violating their federal waste-water-discharge permits.[28] One goal of the suit was to put pressure on the government to enforce the law.

But while some have found the law a useful tool for combating environmental hazards, others are more skeptical. As the case history from the Coalition for a United Elizabeth illustrates, lawsuits have both political and financial costs. First, as even the success stories indicate, lawsuits can drag on for years. A labor activist wryly observed, "If you're a worker or a community person and you go to court with a lawsuit, you have a fifty-fifty chance of living till the case gets decided." When a hazard is endangering lives *now*, it is hard to depend on a protective process that takes so long. Further, lawsuits consume money as well as time. Scott Reed, an environmental lawyer, noted that "litigation budgets are like Department of Defense budgets. The costs are inevitably underestimated."[29] Even the simplest legal effort can cost several thousand dollars, particularly if expert testimony is involved.

Some groups have sought to minimize their costs by sharing lawyers with other organizations or by assigning nonprofessional tasks to law students and volunteers. Others have found lawyers who will work on a contingency basis, receiving pay only if the case is won. For instance, lawyers representing the Hardeman County, Tennessee, residents who were suing Velsicol Chemical Company for $2.5 billion agreed to accept one-third of any settlement in payment.[30] Nevertheless, even a contingency fee will seldom cover court costs, laboratory analyses, expert testimony, and so on.

As the CUE organizers noted, a further problem is that the time and money invested in court cases are not available for other important activities, including community education and organizing. Once a case develops, it takes on its own momentum, and it becomes difficult to devote anything less than a total effort to winning it. The group's original priorities can easily be lost, particularly if lawyers come to make more and more organizational decisions. As the proceedings drag on, the political mobilization that led to the suit can wither and die.

Some groups find that filing a lawsuit can dramatically change the alignment of political forces in their area. Frustrated by the difficulty of getting action on a contaminated city landfill, a Staten Island group sued the city government. The mayor immediately ordered city agencies to end all cooperation with the group. Officials who had previously been helpful in giving information and defining the scope of the problem were unwilling to defy the mayor, and the group lost some valuable allies. The residents might have chosen to take legal action anyway, but their choice of a legal rather than a political strategy reduced the chance for quick remediation and provoked a confrontation they had not expected.

Others go further and doubt the effectiveness of legal action under almost any circumstance. David Wilson, a professor of chemistry at Vanderbilt University and a frequent consultant to community environmental groups, puts it this way:

> America has the best courts money can buy, but most of us are too poor to buy them. Our opponents do not generally suffer from this handicap. It is therefore to our advantage to try our cases in the court of public opinion, where justice is usually cheap and prompt, the penalties assessed against the guilty are certain and severe, and from whose judgment there are no appeals. Don't waste your time and money in court if you can possibly avoid it, because you'll almost surely lose.[31]

That major corporations continue to break the law despite the spate of lawsuits lends weight to Wilson's argument. Hooker Chemical Company, for example, has 1,300 suits pending against it, with claims involving billions of dollars, yet their environmental practices continue to generate opposition.[32] Johns-Manville

Corporation responded to the 16,500 claims by workers contending that the company had inadequately protected them against asbestos by filing for bankruptcy, an act that protects a corporation from its creditors and their suits.[33] Corporations know that their lawyers and lobbyists can usually outspend and outwait legal actions initiated by community groups operating on shoestring budgets. And as the Johns-Manville case shows, when the costs of inadequate protection go up, corporations simply change the rules of the game to protect their profits.

Working in the Electoral Arena

Many environmental groups have moved into the political arena in order to pursue their goals. The case histories described two campaigns to pass a ballot initiative—a law that voters propose by petition, then approve or reject at the polls. Another method is a referendum, which asks voters to ratify or reverse an act of the local or state legislature.

In the first decades of this century ballot initiatives and referenda became popular in the western part of the United States as a way to challenge unresponsive or corrupt legislatures. They were used to pass a hodgepodge of progressive and conservative measures, ranging from woman suffrage to alcohol prohibition. By the early 1970s environmentalists had discovered that ballot initiatives could be used as a tool to fight against such things as nuclear power, toxic waste disposal facilities, and pesticide spraying, and for recycling glass, paper, and metals. By 1981 twenty-five of the fifty states had provisions for initiatives or referenda. Sponsors were required to collect signatures endorsing the measure from a certain percentage (anywhere from 3 to 15 percent) of the voters in the most recent gubernatorial election. Ballot initiatives had two main goals: to enact new rules (e.g., for licensing nuclear power plants) and to make policy statements (e.g., in support of the nuclear freeze). There were a number of successes using this strategy. In 1978 in Montana, for example, voters passed a measure stating that once a proposed nuclear power plant met certain rigorous safety critieria, the plant would then have to be approved by the Montana electorate.[34] In that same year residents in Kern County, California, voted on a special ballot against building a nuclear power plant in the San Joaquin valley.[35]

In 1977 thirty-eight Vermont communities used their town meetings to approve restrictions or bans on the transportation of radioactive materials through their neighborhoods.[36]

A second strategy is to seek to elect candidates who support environmentalist views, or defeat those who oppose them. By campaigning for or endorsing an official who promises to work for specific environmental measures, a group tries to place allies in government offices where they can help it to win its goals.

In the last few national elections, environmental organizations have played an increasingly important role in helping candidates with good environmental records to win seats in Congress. In 1982, for example, several pro-environmental legislators retained their seats due in part to aggressive campaigning by activists. Conversely, environmentalists helped to retire a few congressmen who had opposed stricter regulation.[37]

But it is primarily at the local level that community organizations have used the electoral process to correct specific environmental health hazards. In Heard County, Georgia, for example, Georgia 2000, a local group that opposed a proposed toxic waste facility, ran one of its members for the county commission. After he won, he supported legislative efforts that eventually succeeded in preventing construction of the facility. "The people of Heard County elected me to keep the dump out," he proclaimed. "They'd run me out of town if a deal were ever made."[38]

A member of the Toxic Chemical Task Force in Santa Monica, California, reported that "we elected a city council sympathetic to our aims and they then passed our ordinance that required the registration of toxic materials." Environmental groups have sometimes worked out an ongoing relationship with elected officials. An organizer from an antitoxic materials coalition in California, for instance, noted that the "mayor is a member of our group. We helped him and other concerned officials with their campaigns."

Sometimes, however, the relationship with an elected official is adversarial. A New Jersey group explained that their constituents "replaced those officials who opposed" their efforts to prevent Dow Chemical Company from building a "storage farm" for toxic chemicals. "If politicians tried to work against us," said one member, "we generally played hardball politics in return."

Rather than sponsor a particular politician, other groups have developed a platform or position paper. Candidates who endorse the statement are supported; those who refuse are opposed. Many citizen action groups, including the Association of Community Organizations for Reform Now (ACORN) and Fair Share, have used this tactic.

A third strategy is to propose, advocate, and lobby for legislative changes that will reduce or eliminate an environmental health hazard. Since such hazards often result from inadequate or poorly enforced regulations, some groups fight for stricter regulatory codes. Others lobby legislators to enact resolutions supporting their aims, urge them to use their oversight powers to investigate other government agencies, or attempt to block legislation that might damage the public's health.

Although these three electoral strategies differ in their specific goals and tactics, they share certain important characteristics. Choosing such a strategy immediately defines the issue as a political problem, rather than a technical or scientific one, thus inviting popular participation in the campaign. Widespread public acceptance of the legitimacy of electoral politics further enhances the appeal of these strategies. On the other hand, some groups believe that the existing political system is itself an important cause of environmental problems; they oppose electoral efforts that legitimate inequitable and unresponsive institutions.

Those on both sides of this debate cite evidence to support their views. On the positive side, a victorious electoral campaign, like a successful lawsuit, can result in the real reduction of a health hazard. The restrictions on nuclear power plants, the bans on the transportation of radioactive materials, and the prohibitions of aerial spraying of herbicides mean that fewer people are exposed to substances that can cause injury or death. Since new legislation or initiatives generally set ground rules for specific hazards as well as for other similar situations, victories can help prevent future problems: the congressional testimony of Love Canal residents in support of stricter control of toxic waste, for example, illustrates how one group can contribute to laws that help others.

On another level, victories in the electoral arena have won important new rights. The Philadelphia right-to-know law gives residents and workers the right to get information on hazardous materials in their community or workplace. Such knowledge is an important prerequisite for effective protective action. A Montana ballot initiative that gave voters in that state the power to approve or reject construction of new nuclear power plants expanded the democratic rights of Montanans. Electoral efforts that win new power for people can provide a lasting tool with which to fight for change.

Victories by community environmental groups also have important side-effects. A Long Island group that forced the town government to pass an ordinance banning the dumping of toxic chemicals later persuaded the town to take violators to court. The town thereby had to provide for the court battle. The member of Georgia 2000 in Heard County who was elected county commissioner used that position to influence other commissioners to support the fight against the proposed toxic waste facility. With the county commissioners supporting its aims, Georgia 2000 could focus its energy on winning other allies. Folks Organized for Responsible Energy, a Minnesota group fighting uranium mining, introduced a bill to fund a citizen board to investigate applications for mining permits—allowing them to put their energy elsewhere.

As important as the concrete victories are the opportunities electoral work creates for public education. The Delaware Valley Toxics Coalition's (DVTC) educational campaign on the right-to-know bill, for example, alerted the population of Philadelphia to the toxic chemical peril it faced. Similarly, Massachusetts Fair Share's canvassing in support of its legislative proposals on toxic wastes allowed the group to talk to thousands of residents.

Still another benefit of electoral campaigns is their contribution to the growth and development of activists and their organizations. Elizabeth Davenport, the coordinator of Environmental Action, a national organization that played an active role in the 1982 elections, observed that a political campaign "gets people trained, gets them politically active and gets them ready for future issues and campaigns."[39] Organizations also prosper. For example, after the 1978 antinuclear initiative in Montana was passed, its sponsoring group suddenly mushroomed from "a demoralized and factionalized group of four petitioners into a statewide organization."[40] Four autonomous local groups sprouted out of the New Jersey campaign in support of the ban on uranium mining.

Electoral victories breed further success. Caron Chess of the DVTC noted that "before the bill was introduced we had certain people involved. When it was introduced we had more. When it appeared we were going to win, we'd get people saying things like: 'I didn't think right to know had a chance, but I'm going to help now because you really might make it.' Now when people think of toxics problems in Philadelphia, they often think of DVTC because it's the organization with the highest profile."[41]

Finally, electoral campaigns provide a useful focus for building alliances and coalitions. In Mendocino County, California, for example, the initiative against herbicide spraying, described earlier, brought together environmental activists, counterculture youth, apple growers and other small farmers, and medical personnel. For organizations that hope to reach beyond their obvious constituencies, electoral work has often provided a useful starting point for building such alliances. The political legitimacy of work in the electoral arena often makes it easy to draw in elements who might be unlikely to participate in more controversial strategies.

Opponents of electoral strategies point out their risks and limits. Like a lawsuit, electoral work can divert an organization from its primary goals. Initially, most community groups decide to enter the political arena to remedy a particular problem. Once the campaign gains momentum, however, it is easy to lose sight of the relationship between ends and means. For example, a member of a Wisconsin group that won a nonbinding countywide referendum restricting nuclear waste disposal reported that, despite the referendum's success, "there was no long lasting effort or self-propagating energy" developed from the campaign. After the successful vote, he said, "much steam was lost."

Victories won in the electoral arena can prove to be ephemeral. For instance, the Washington Heights Health Action Project worked for a year to convince the New York City Council to pass a bill strengthening the city's ability to prevent lead poisoning in children. But the group lacked the power to ensure that the new law was enforced. While it is relatively easy to get a politician to promise to support a community group's demands, it is more difficult to translate these promises into meaningful action. As one New Jersey activist said, the politicians " 'yes' you to death and then they don't do anything." And as we have seen, industry has developed numerous ways to influence legislators. By aggressive lobbying, campaign contributions, or even bribes, corporations can often thwart even the most determined community group.

Activists have developed a variety of ways to hold on to promises of support or to actual victories. These include requesting written statements of support from candidates, establishing ongoing task forces or other semiofficial bodies with strong citizen representation to monitor enforcement of new laws, and carrying out public education campaigns to expose corporate influence on legislators. But for some, involvement in the electoral arena has been disillusioning. Edwinda Cosgriff of BLAST described her own conclusions, based on a decade of experience: "I rarely believe anything a politician says. I'm very cynical. I realize they have to lie, compromise and make deals. They're not that concerned with the individual voter, they're more concerned with the people who contribute to their campaigns. Only when you get *numbers* of people—then they become worried."[42]

Obviously, when environmentalists attempt to thwart a corporate project they will meet with opposition. And almost always industry will have more money and other resources than will a community group. For these reasons, some activists believe that choosing to battle in the electoral system automatically concedes two important advantages to their opponents.

First, the rules of electoral politics are fixed. Both sides adhere to the same timetable, both have to meet the same requirements. Electoral work forces community activists to relinquish the one advantage they otherwise have: the ability to use surprise tactics to make up for few resources. Second, major polluters generally have more experience than do environmental groups in working within established political institutions. They have helped to set the rules and choose the players; they know the power brokers and their patterns of influence. They are therefore usually the more skillful players.

On occasion environmental groups have been able to overcome these disadvantages: for instance, the Montana nuclear industry lost, despite the fact that it outspent the supporters of the ballot initiative by twenty-six to one.[43] But for every success there are many failures. A 1979 study by the Council on Economic Priorities, a New York research group, showed that the side spending the most won eleven of the fifteen ballot initiatives investigated.[44] Although we might like to believe the opposite, Goliath wins more often than David. Moreover, had David chosen to slug it out, rather than surprise his opponent with a slingshot, his loss would be forgotten.

Some community organizations find that avoiding the established political system provides them with a mantle of moral credibility that allows them to make greater demands on the community's time and energy than a more established group would be able to do. The widespread conviction that politics is corrupt and dirty can taint a group that enters the electoral arena. Carol Froelich explained how this perception led We Who Care to reject the mayor of Rutherford's request that their investigation of childhood leukemia be conducted by the city government: "There was a lot of suspicion among people that anything involved in politics wouldn't be honest, that we still might not be able to find truth in what the state experts said, so we decided to stay separate."[45]

Critics of electoral work also charge that it can lead to a dilution of political principles—compromises are inevitable if a majority coalition is to be built that will pass an initiative, elect a candidate, or approve a bill. The question is not one of moral purity—probably an unattainable goal these days—but of the danger of losing sight of the broader vision that motivates political activity. Thus, for example, national environmental groups that have campaigned for candidates with good environmental records have found themselves endorsing people with unacceptable positions on defense spending and social issues. For those with a more radical perspective on the political changes necessary in this country, supporting a friend of a strong clean air act who also votes for the MX missile and restrictions on government funding of abortion may be an unpalatable compromise.

In addition, such compromises affect a group's political support. When environmentalists ask black or women's groups for help in a particular struggle, they should not be surprised if aid is withheld from those who have worked to elect a politician who opposes women's or black issues. Unless a group is clear as to its political principles, the electoral arena can be a dangerous place to look for bedfellows.

Finally, some people believe that electoral politics reinforces a passive notion of how to bring about change. The U.S. public tends to see the established political system as "a marketplace of buyers (voters/members) and sellers (politicians, leaders, parties and platforms)."[46] Voters look for perfect candidates or platforms, then turn over to them the responsibility for solving the problem. Mike Miller, a veteran community organizer, argues that instead "citizens ought to be active, not reactive. They ought to formulate as well as respond to options."[47] Electoral politics can minimize this active role by relinquishing the ultimate responsibility for enactment or enforcement to government, an institution especially vulnerable to corporate influence. Unless environmental activists are clear about the limits of electoral politics, choosing this strategy can reinforce the dominant belief that change in the United States comes only through the ballot box. If an electoral effort fails, it then becomes harder to convince people of the possibility and necessity of other strategies for change.

Guidelines for Legal and Electoral Work

Obviously, groups decide which strategy to use on the basis of their available resources and on their analysis of the specific situation in the community. But the following guidelines may help to inform these choices.

First, most successful groups pursue several strategies simultaneously. For example, at the same time that Georgia 2000 helped to elect a county commissioner, some members were demonstrating and collecting signatures on a petition against the proposed toxic waste facility, while others guarded the site with rifles and shotguns to prevent construction from commencing.[48] Clearly the outcome did not depend on an electoral victory alone. Similarly, Gregor MacGregor, an attorney who has represented many New England environmental groups, has noted that "rarely is an environmental lawsuit won when there is not a political constituency built to accompany it."[49] By pursuing many strategies, a group decreases the likelihood that its opponents will be able to smash its one basket of eggs in a single swoop, and it also teaches its constituents that victory requires a broad attack on several fronts. Because legal and electoral battles take place on terrain where corporate interests have a dominant voice, it is particularly important to combine them with other strategies such as community education and organizing.

Second, groups must consider the broader political context in which they choose their strategies. In the early 1970s, when the environmental movement was riding a wave of popularity, judges, juries, and legislators were more likely to consider environmental initiatives favorably. Now, with industry and the Reagan administration fervently attacking environmental regulation, no judge or lawmaker is immune to their cost-benefit rhetoric. The early 1980s might *not* be the time to look for justice in the courts; rather, using other political strategies until the judiciary takes on a more progressive tone may be more successful. Similarly, some environmental groups advocate state or local, rather than national, electoral work until the political climate in Washington improves. On the other hand, external political factors can improve a group's prospects or provide unexpected dividends. Using lawsuits as a delaying tactic, as harassment, or as a bargaining point may make sense if an environmental group has other political strengths or expects a realignment of political forces in its favor. The antinuclear power movement effectively used legal tactics that, in conjunction with the industry's economic problems and mass demonstrations, slowed the growth of nuclear energy.

A third suggestion is to involve all group members, not just the leaders, in legal or electoral efforts. For example, the membership should discuss and vote on lawsuits before they are filed and those who are handling the case should report on its progress at regular intervals. Organizing a campaign to pass a bill or referendum can involve new and old members in lobbying, testifying at public hearings, distributing leaflets, writing letters to legislators, demonstrating, and so on—all actions that increase members' commitment to the group and its cause. Active participation also helps to prevent lawyers or legislative aides from dominating the group, a fate that can afflict those who allow a court case or bill to become an end in itself rather than a means to a healthier environment.

Since legal and electoral work requires major commitments of time, energy, and money, groups often look for allies who can share or even take over this burden. Sometimes political pressure can force the government, rather than a citizens' group, to initiate legal action; in other cases, national environmental organizations (the Environmental Defense Fund, the Natural Resources Defense Council, or the Sierra Club) can be persuaded to have their experienced legal staffs take on an issue. The local group can then join the suit either as an *amicus curiae*, or friend of the court, a position with few formal responsibilities, or as an intervenor, a status that, according to environmental lawyer Scott Reed, "allows many of the benefits of litigation without the full costs and exposure of being an instigator."[50] On the legislative front, community groups have also successfully turned over primary responsibility to organizations such as the League of Women Voters or the Public Interest Research Group. Such a delegation of tasks permits the local group to focus its energies on other political strategies.

Finally, legal and electoral work contributes most to building a movement when it is used to win concessions that expand environmentalists' power and influence within the political arena. Court decisions that set a precedent for similar cases, damage suits that win dollars for community groups, right-to-know laws, the establishment of citizens' monitoring boards and referenda that allow a community to refuse a nuclear power plant or a toxic waste facility, all give people the power, limited as it may be, to contest future decisions that may threaten them. Moreover, such victories create new governmental forums in which environmentalists can present their case. Public education on these measures addresses not only the hazard and its consequences but also who has power and how they use it. And by winning new power for themselves and for the people as a whole, environmental groups avoid the problem of becoming wedded to a particular agency's or administration's commitment to environmental protection.

8. STRATEGIES FOR ACTION:
Community Organizing

Case History 1: Fort Worth, Texas

The Association of Community Organizations for Reform Now (ACORN) is a national organization of low-income and working people that has local chapters in twenty states across the country.[1] ACORN organizes community residents to demand changes in their living conditions and in government and corporate policies. In this case history, an ACORN organizer describes a 1980 campaign against a company polluting the air in a Fort Worth, Texas, neighborhood.

The Fort Worth ACORN staff had its eye on the Stauffer chemical plant for over a year. In July a new organizer came here and she was assigned to the area where the plant is located. People complained of the fumes, and we discovered that the history of relations between the plant and the neighborhood was a stormy one. About twenty years ago, the plant experienced a severe upset, releasing a huge belch of sulfuric acid mist and literally wiping the paint off dozens of homes overnight. Stauffer settled out of court with a number of homeowners over the property damage. Health effects don't seem to have been an issue at the time.

Then ten years ago the company installed three continuous sulfur dioxide monitors in the neighborhood. Sulfur dioxide (SO_2) is the major pollutant produced by the plant, a by-product of their sulfuric-acid manufacturing process. The machines were there documenting the fact that the plant was in compliance with existing regulations, but in recent years, the neighbors became convinced the fumes were worse.

The new organizer decided to make air pollution her issue and organized a couple of meetings. The first was with the head of the city health department's air pollution control division. He turned out to be an ally in the campaign, but not an open one. His department's resources were very limited, and it would have been political suicide for him to have been identified with us. But he supplied us with valuable tips and information throughout the campaign. From him, we learned that (1) Stauffer had installed and maintained the SO_2 monitors on their own, although they were under no legal obligation to do so; (2) neither the city nor the state had the personnel to maintain continuous monitoring equipment at the plant; and (3) the company was 'getting away with murder,' but the health department couldn't document it.

After that, the organizer set up a meeting with the plant manager. He came on sweet, fetching coffee for everybody, showed the ACORN members the graphs from the monitoring machines, and so on. It was a bust, but it seemed to worry Stauffer that folks were looking into it. The organizer then decided to conduct a health survey of the residents, to see if statistics would show that their health was being affected.

In the midst of the survey, the organizer left, and the campaign was almost derailed permanently. It had been entrusted to a couple of new leaders who were less than dependable. The upshot was that we wasted about a month before we scrapped the whole idea of a survey.

Meanwhile, after our meeting, the health department's air pollution official had asked the State Air Control Board to come out and inspect the company's monitors. He passed a copy of the report to us before it went to the company. It was a real shocker—none of Stauffer's monitoring devices worked. So we decided to go after new monitors, with secondary demands of verifiable, city-supervised maintenance and a list of established procedures to follow in the event of higher than normal emissions. Although we had no legal handle on the monitors—the law doesn't require them—we thought the precedent of Stauffer installing them the first time, along with the disgraceful Air Control Board report, gave us the leverage we needed.

We invited the district's city councilman to a negotiating session as a mediator, and after he had accepted, we invited Stauffer. Under the circumstances, they thought it best to attend. Meanwhile, some guy in the neighborhood—not an ACORN member—called up a local TV station and screamed bloody murder about the plant. A young would-be video crusader from the station came out with a camera and filmed steam coming out of the plant (it was just water—SO_2 is invisible). The reporter identified these vapors as deadly and accused the plant of systematic genocide in the name of profit. What disturbed us was that the TV station used our name to gain entry to some of our sources, never even spoke to us, and created the impression that we had sparked their investigation. It was a real hatchet job and the plant manager called us up screaming.

We told him he'd been had, and that when we decided to bring the press in, we'd be down at the factory gates with our members and our banner. He decided he better deal with us before he was slandered on the tube again. We promised him one shot at negotiating with us and the councilman, with no press present.

The negotiating session was a piece of cake. We had a list of EPA-approved equipment, and a letter to our members informing them of our win, with a place for the manager to sign. Stauffer flew in their vice-president for public relations from Houston, who delivered a heartwarming speech on corporate responsibility, and then signed. A week later they took us on a tour to show us the new devices and gave us a maintenance manual. They bought four of the monitors, not three, at about $18,000 apiece.

We started on another campaign in that neighborhood but now we're back to Stauffer. They had another big emission, while the manager was out of the country, and the plant employees on duty didn't follow any of the guidelines. They didn't shut down or call the health department until we had done so ourselves. So now we have a campaign around the emergency procedures.

Case History 2: Warren County, North Carolina

On a summer night in August 1978 a tanker truck drove slowly along Route 210 in central North Carolina, releasing its cargo of oil as it went.[2] Soon residents along the highway began to complain of foul odors—"like rotten eggs, ammonia and vinegar," as one woman put it—and eye, ear, nose, and throat irritations.[3] An investigation by the state identified the substance as waste oil contaminated with polychlorinated biphenyls (PCBs), the toxic chemical that can cause severe skin rashes and birth defects, and found traces of it along more than two hundred miles of roadside. The state hand-delivered four thousand letters to residents along the highway, warning them not to eat or sell produce or livestock grown within one hundred yards of the road.[4]

A month later the North Carolina Bureau of Criminal Investigation arrested a New York businessman and his two sons, charging them with the illegal midnight dumping. The three pled guilty and agreed to serve as state's witnesses. They testified that the operator of the Ward Transformer Company of Raleigh, North Carolina, had paid them $75,000 to get rid of 31,000 gallons of PCB-contaminated transformer oil. The state than filed a $12.5 million civil suit against Ward and the truckers.[5]

Late in 1978 North Carolina Governor James Hunt announced that the state was planning to dispose of the PCB-tainted soil at a new dump site in Warren County, in the northeastern part of the state, despite the fact that the water table at the proposed site was within fifteen feet of the bottom of the dump, considerably less than the fifty feet the Environmental Protection Agency (EPA) recommends between water and waste.

Why did Governor Hunt plan to dump the PCBs in such an unsuitable location? The newly formed Warren County Citizens Concerned About PCBs charged that their county had been chosen because it had little political power: its population is sixty-four percent black and American Indian and its median income is the lowest in the state.[6] Others believed the choice was simply racist; as Leon White, a black minister and civil rights activist, put it, "They chose a black community [because] they want to experiment on our people."[7] Angry residents also feared that Hunt wanted to make this the first of a series of dump sites to accept hazardous wastes: North Carolina is the nation's eleventh largest producer of toxic wastes and Warren County is only fifty miles from a heavy concentration of chemical industries.

For almost three years the county government and its residents sought to block the proposed dump through legal action. Every effort failed and, in May 1982, the EPA and the state agreed to provide $2.5 million to construct the dump site for the PCB wastes. The Citizens Concerned About PCBs then decided to act more vigorously. On September 15, 1982, about 125 demonstrators marched to the entrance of the landfill in order to block the first delivery of toxic soil. Leading the march were Leon White and Ken Ferrucio, a white community college teacher and president of the group. Chanting slogans such as "Dump Hunt in the Dump" and carrying placards reading "Oh Lord, don't let 'em drop that PCB on me," the demonstrators were met by a squadron of sixty riot-equipped North Carolina highway patrol officers. Fifty-five marchers were arrested that day and bused to the county seat. Governor Hunt vowed he would not meet with the protestors until the landfill operation was completed.

For two weeks demonstrators continued to try to block the trucks. Eventually 520 people, including 100 children, were arrested, including Reverend Joseph Lowerey, president of the Southern Christian Leadership Conference, and Walter Fauntroy, the District of Columbia's delegate to the U.S. Congress.[8] Their commitment to nonviolence helped to prevent a single injury.

Many others participated in the struggle. Community meetings attracted more than a thousand people. A support rally at the University of North Carolina–Chapel Hill drew two hundred students, some of whom then traveled to Warren County to picket.

Within the county, blacks and whites made some first steps toward breaking down years of segregated political activity. As Ken Ferrucio put it, "We marched together, prayed together and struggled together, so as a result some very close human relationships formed."[9]

In October, after publicity had brought the issue national attention, Governor Hunt finally decided to meet with the demonstrators. He met with eleven protest leaders, and promised to support state legislation prohibiting future landfills in Warren County.[10] While local activists were skeptical, the governor's pledge did show that they had made the PCBs a major political issue for blacks and whites. Ferrucio explained the community's accomplishment: "The toxic waste question is now linked inextricably to the human and civil rights issue. If we've done nothing else, we've made that link."[11]

Almost every local environmental group organizes people in the community. But while every pamphlet, letter-writing campaign, public meeting, and demonstration uses techniques developed by community organizers, in this section, community organizing will refer only to those cases in which a community group directly confronts a government agency or corporation in order to force it to end practices suspected of being hazardous. Unlike legal or electoral work, community organizing does not look to a third party (the courts or the legislature) to intervene but chooses as its primary target those with the power to grant a group's demands. It seeks to change the power relations between ordinary people and their opponents. Community organizers use a variety of tactics, including educational campaigns, demonstrations, public meetings, and civil disobedience. No matter what the tactic, however, the goal is to muster enough political clout to force the opponent to give in to the group's demands.

"Just as a lens can concentrate the rays of the sun to start a fire," explains Heather Booth, a trainer of organizers, "a direct action organization concentrates the power people have in order to win a victory."[12] In Forth Worth, Texas, for example, community pressure led Stauffer to install new air pollution monitoring devices. In Warren County, North Carolina, although mass arrests did not stop the governor from dumping PCB-contaminated soil in a landfill, they did make the problem a national issue, and led to a promise not to build more dump sites in the county.

In the past decade multi-issue citizen action groups have sprung up across the country. Their goals have been to build community organizations that could represent poor and working people in their fight for better living conditions, and they have tackled such issues as housing, taxes, energy rates, and social service cutbacks. While many of these groups are locally based (like the Coalition for a United Elizabeth), others, such as ACORN, the Citizen's Alliance, Citizen Labor Energy Coalition, and Fair Share, have built statewide, regional, or even national organizations. Recently, many of these groups have begun to take on environmental health issues. Massachusetts Fair Share, for instance, which has more than thirty member-run chapters throughout that state and a membership of more than 60,000, explains why the group believes that community organizing is the best way to improve environmental conditions:

> We cannot wait for city councils, the state legislature or congress to act for us, or pass laws that will empower citizens to address this crucial problem [of toxic chemicals]. We must exercise these rights on a case by case basis, to create a groundswell of citizen support and action which will force improved laws, regulation, and enforcement.[13]

Massachusetts Fair Share's Health and Safety Project organizes people to win three basic rights: (1) the right *to know* what chemicals and wastes people are being exposed to in their neighborhoods and workplaces; (2) the right *to inspect* industries, treatment facilities, and landfills that use and process toxic chemicals and wastes; and (3) the right *to negotiate* directly with corporations and dump-site operators for the elimination of hazards discovered through inspection and research.[14] Fair Share organizers use a variety of tactics to win these rights, from demonstrations, to supporting political candidates who endorse the group's demands, to pressuring local agencies to enforce existing laws. Should a corporation refuse to allow a community inspection, Fair Share will call in what it calls "an arsenal of agencies" to inspect the company's compliance with fire regulations, building codes, and occupational safety and health and environmental regulations. If necessary, members will picket each government agency until it carries out its mandated inspection. Many companies have decided it is easier to accede to the group's demands than to face a parade of inspectors.

Fair Share has several victories to its credit. In Lowell, Massachusetts, for example, the Ayer City Fair Share chapter demanded that the state clean up 68,000 gallons of hazardous waste left at an abandoned dump site by the Silresim Chemical Corporation. After demonstrations, meetings with public officials, and extensive press coverage, the state agreed to the group's demands.[15] Other chapters have won the right to inspect factories in their neighborhoods and some companies have reduced pollutants on the basis of the group's recommendation.[16]

In other parts of the country citizen action groups have used direct action to win other sorts of improvements in health conditions. A demonstration and sit-in near a General Electric warehouse storing PCBs in a Philadelphia neighborhood convinced the city government to order the warehouse closed and to post police to make sure no more PCBs were brought in.[17] An ACORN chapter in Memphis, Tennessee, organized seventy people to march to a chemical plant to demand that its pesticide production unit be moved out of their residential neighborhood. Since pesticides were only one of the company's products, only a few jobs were lost when it gave in to ACORN's demands.

But at other times the activities of citizen action groups have not brought about change. Then, a community fearing that its very survival is at stake may choose to use civil disobedience or extralegal tactics such as sit-ins, blockades, or even sabotage. The mass arrests in Warren County, North Carolina; the confrontations over nuclear power in Seabrook, New Hampshire, and Diablo Canyon, California; and the detaining of EPA officials at Love

Canal illustrate the potential for militant action within the environmental movement.

The decision to break the law usually comes after all other strategies have failed. In Minnesota, farmers sought for years to block a high-voltage power line that was to carry electricity from a generating plant in North Dakota to a substation outside the twin cities in Minnesota.[18] They opposed the high-voltage lines because the state could use its power of eminent domain to seize their land for the project and because they feared the electric current could damage their health or harm their livestock. From 1973 to 1977, farmers used legislative lobbying, lawsuits, and community organizing in an attempt to prevent the power line from going up, but their efforts failed.

In June 1976, as surveyors began to make measurements for power line construction, Virgil Fuchs, a Stearns County dairy and crop farmer, drove his tractor toward the surveyors, smashed their tripod, then rammed his tractor into the utility company's pickup truck. At a court hearing in St. Cloud the next day, a crowd of farmers from throughout Western Minnesota gathered to protest Fuchs's arrest. When Fuchs refused to post bail, officials released him rather than provoke the irate farmers.[19]

Beginning in 1976, a posse of fifty or sixty farmers using CB radios confronted surveying teams, obstructing their work. These skirmishes continued for more than a year. Early in 1978 eight farmers were arrested but rather than prosecute, the local district attorney resigned. "They're not criminals," he said, "and that's the point. The law is clear and they are breaking the law as far as I can see, but I don't agree with the law."[20] No local attorney could be found to replace him. But the state legislature refused to enact a moratorium on the power line, and so protests became increasingly violent. In February 1978 a pitched battle broke out in Stearns County, where more than two hundred farmers armed with baseball bats used manure spreaders, tanks of anhydrous ammonia, and tractors to confront surveyors and the one hundred state troopers protecting them.[21]

In the following weeks more farmers were arrested during peaceful civil disobedience. More than eight thousand people joined a "March for Justice" in Pope County, Minnesota, which was described as follows in a book on the power-line struggle: "Following a tractor with an American flag, a coffin labeled 'Justice,' and an enormous papier-mache figure, the 'Corporate Giant,' the people marched solemnly from Lowry to Glenwood to demonstrate their opposition to the power-line and their support of the protest."[22]

Eventually the power line was built, but local opposition did not end. As the power company completed stringing its wires across Pope County, towers began to fall: nearly every night "bolt weevils" came out to remove bolts from the towers, causing them to collapse. The power company increased security and offered a $100,000 reward for information leading to the arrest and conviction of any tower toppler.[23] But while politicians and the media condemned the destruction of property, local residents were more supportive. Patty Kakac, a young woman who lived on a farm in Douglas County, described her reaction this way:

> It's funny. Several years ago I would have thought pulling towers down was extreme. Not now. I am almost to the point where it is not extreme enough. It is real sad. I wish it wouldn't take this. But people wouldn't pay attention to the struggle that the farmer goes through until something drastic brings their attention to it. As soon as a tower goes down, almost the whole state knows about it. Maybe it puts a question in people's minds—What is going on? What is happening here?[24]

Another resident, Dick Hanson, a farmer and a political activist, was even more forceful in his justification of the toppling.

> Many different people support the sabotage and I think it is because of the rough way the power companies along with the government rammed this thing down our throats. They had the courts working for them, they had the money and power to push all these things through. Now we have the power. They cannot put a security guard behind every cornstalk and every tree and under every tower twenty-four hours a day. It certainly is unusual, but I am not so sure it is out of the ordinary. Take the Boston Tea Party incident or a heck of a lot of other incidents where people had to take things into their own hands because there was no legitimate avenue open. That is the case now.[25]

The Love Canal residents got a quick response to their extralegal acts. Two days after the federal government released the study showing that eleven of the thirty-six residents tested had some chromosomal abnormalities, an angry crowd of more than three hundred people "detained" two Environmental Protection Agency (EPA) officials in the Love Canal Homeowners' Association office for more than five hours, demanding immediate evacuation from their polluted neighborhood. The FBI designated the incident "a hostage situation" and threatened to disperse the crowd and arrest the "kidnappers," who served their captives cookies and sandwiches. Rather than provoke a violent confrontation, the Homeowners' Association decided to release the officials. Lois Gibbs told a reporter that the incident was "a Sesame Street picnic compared to what might happen" if the government did not evacuate the neighborhood and reimburse residents for their losses.[26]

Two days later President Carter designated Love Canal a federal emergency area, clearing the way for the relocation of more than seven hundred families. Gibbs observed that as a result of the action "we've gotten more attention [from the White House] in half a day than we've gotten in two years."[27]

Some community groups have tried to win concessions by using economic pressure. The Long Island Safe Energy Coalition, for example, urged ratepayers not to pay rate increases demanded by Long Island Lighting Company in order to pay for the Shoreham nuclear power plant. Two hundred seventy-five families pledged not to pay any increase as long as one thousand other families would do the same. This target was not achieved and the group dropped this tactic. However, the threat did receive excellent publicity. Similarly, a Massachusetts Fair Share chapter worked on a campaign to force a bank that loaned money to a toxic-dump-site operator to contribute to the cleanup. Organizers convinced both institutions and individuals to pledge to withdraw their accounts should the bank refuse to help remove the hazard. Tennesseans Against Chemical Hazards, a coalition of community groups, explored the possibility of a consumer boycott of the products of the Velsicol Chemical Company, a company whose pollutants several local groups were battling.[28] Eventually the group rejected a boycott, deciding that it was too indirect a tactic to affect a multinational corporation.

National labor, church, and peace groups have effectively used stockholders' resolutions to expose the social consequences of corporate policies. In a stockholders' resolution, individuals or institutions owning shares in a company request information or propose changes at the annual meeting of shareholders. Resolutions that are approved by more than 3 percent of stock owners can be resubmitted the following year. While few shareholders' resolutions pass, they are a useful tool for public education and debate on corporate behavior. In cooperation with national organizations—especially church groups with large stock portfolios—community groups can use this tool to put additional pressure on a polluting company.

In sum, community groups have generally not been effective in applying economic pressure on corporations. Whether a coordinated national strategy for this kind of tactic could be successful has yet to be tested.

In the face of weakening government regulation of environmental hazards and lax enforcement of existing laws, community organizing becomes a particularly useful strategy. The Massachusetts Fair Share toxics handbook asserts that "through direct activity neighbors armed with technical and scientific information . . . have a much better chance to guarantee health and safety than the understaffed, underfunded and uncoordinated local, state and federal agencies."[29] As opposed to the politicians, judges, and government officials who cave in to industry's environmental blackmail, militant community groups can insist that health and safety concerns are addressed in the political arena. And as Frances Piven and Richard Cloward have argued in their book *Poor People's Movements: Why They Succeed, How They Fail,* protest movements are most effective when they use their power to disrupt the smooth functioning of the system, which is precisely the outcome of community organizing and direct action.[30]

As corporations and their conservative supporters organize to defend their interests, judicial or electoral battles sometimes dissipate rather than concentrate the movement's energy and momentum. By confronting local threats to health at their source, environmentalists set the stage for a head-to-head battle between polluters and their victims. Not only does this help to win victories, it also strengthens organizations. In some cases, an individual's actions inspire the group. For example, after the farmer Virgil Fuchs used his tractor to chase surveyors off his Minnesota farm and to destroy their equipment, a neighbor observed, "Virgil is quite a hero around here. Almost everything started with Virgil's running over the equipment. From then, people's involvement in our area became more intense."[31] In other situations, collective action can breathe new life into an organization. Confrontations start the adrenaline flowing and force participants to define their politics and test their commitment, leading to a more dedicated and skilled membership.

Even when action fails, its participants often learn important positive lessons. For instance, John Tripp, another Minnesota farmer, told an interviewer before the power lines went up: "Even if we do lose and get the line I think we have done a tremendous amount of good for farmers in general. So many farmers have told me that, too—that they think it is absolutely remarkable how well farmers have stuck together. Most farmers say it has never happened before in their lifetimes—I mean the older farmers. I think we have showed farmers we can fight these things."[32]

Despite these advantages, community organizing has its limits. Strategies and tactics that work today may fail tomorrow. Wade Rathke, chief organizer for ACORN, explained one dimension of this problem: "As we have increased in sophistication and complexity so have the corporations and government officials we come up against. They are increasingly able to deal with our whole set of tactics quite facilely, leaving some of our tactics hollow."[33] For example, when the state refused to clean up a toxic dump site linked to skin and breathing problems, forty members of the Frayser Health and Safety Committee in Memphis, Tennessee, sat in at the governor's office, demanding a meeting to urge him to declare Frayser a disaster area eligible for federal funds. But the governor ignored their request, ordered fifteen members of the group arrested, and took no remedial action. The action divided the committee and made subsequent organizing more difficult. Similarly, an upstate New York citizens' group demanded a meeting with officials of a corporation to investigate its role in groundwater contamination. At the meeting each resident was greeted by a corporate employee and escorted to a large auditorium where the entire group was fed dinner. The company then made a presentation on its waste disposal processes, at the end of which its spokesman announced with regret that he would not have time to answer any questions and ended the meeting. By ignoring or co-opting community protests, those in power can sometimes deflect opposition. In part, this problem can be avoided by planning events more carefully. In part organizers need to be creative in developing new tactics. But in some cases, a group may simply lack the resources or experience necessary to build enough power at the community level to successfully challenge their opponents. In such situations, a strategy other than local organizing might be more effective.

Almost by definition community organizing, and especially direct action, polarizes people. A member of Ecology Action in Oswego, New York, explained how this conflict can be problematic: "The effect of direct action is very dependent on how the issue is perceived by the general public—mass demos are not good for locally unpopular issues since people coming from outside only antagonize locals, who won't change their minds anyhow. Demos are excellent for popular issues like toxics where you want to show state officials local support."

Since some people benefit from pollution, environmental organizations will probably never have the support of the entire community. But if the tactics of the group create more opposition than the issue itself, victory will be difficult. Illegal acts, especially violence, can turn a community against environmentalists. George Crocker, a Minnesotan long involved in radical causes and an organizer of the power-line struggle, described why the farmers at one point decided against violent confrontation: "Our thinking was, don't set yourself up for being terrorists—law and order is a very strong thing. We gave [the troopers] plastic flowers to show we were not attacking human beings."[34] Later, as we have seen, the power-line battles did erupt into violence and sabotage. The lesson to be learned is that a group's actions must match the level of resistance that a community is willing to offer. If an organization moves too quickly or too militantly, it can cut off the popular support that nourishes it.

Corporate and governmental opponents of community groups will, of course, always believe that any action is too much and too soon. With experience, activists learn to reject both the go-slow advice of their enemies and the more hot-headed suggestions of impatient supporters (or agents provocateurs). The correct criterion by which to judge an action is its ability to strengthen the group and move it closer to achieving its goals.

Community organizing can reflect diverse political perspectives. Some see it as a method for strengthening and empowering people within their own neighborhood. Others regard it as one of many tools for building a national progressive movement. Still others view local organizing as a useful tactic for winning a specific concession. The grass-roots environmental movement has not yet developed a unique theory or practice of community organizing.

Whatever its rationale, community organizing has been, in my opinion, environmental activists' most effective strategy. It has educated millions of Americans about the hazards they face in their own backyards. It has won concrete victories from large and small polluting corporations and government agencies. It has helped to build lasting local organizations that have become forceful advocates for environmental protection. It has demonstrated the popular power and anger that has in turn allowed legal and electoral victories. It has contributed a legacy of militance that will serve as an example for future environmentalists.

The challenges that community organizers face in the years to come are to expand and diversify their arsenal of tactics, to learn how to move from local struggles to regional and national ones, and to choose a level of resistance that can be sustained for a prolonged battle. We will explore how to achieve these goals in the final chapter.

9. COALITION BUILDING:
Issues and Problems—Labor/People of Color/Women/Peace Groups/Third World Groups

Labor

Case History: Oil, Chemical and Atomic Workers' Strike

In January 1973 the Oil, Chemical and Atomic Workers' (OCAW) union called a strike against Shell Oil Company operations in California, Louisiana, Texas, and the state of Washington.[1] Its goal was to force Shell to sign a contract guaranteeing its workers certain health and safety rights by establishing a union-management health and safety committee, hiring outside consultants to inspect the plant periodically, providing medical examinations for workers when plant inspections revealed a problem, and making all company records on illness and death available to the union. Eleven major oil companies had already agreed to these provisions.

But Shell at first refused to negotiate. Management claimed that it alone was "legally responsible for the health and safety of Shell employees in the workplace," and that "this responsibility cannot be shared [with the union]."[2] The union decided to fight. Tony Mazzocchi, then the Washington, D.C., representative of OCAW, insisted "We can't lose this one. This is the first time workers have gone out on what are essentially health and safety issues. We're involved in a pioneer effort."[3] The union called on consumers to boycott Shell products.

It was hardly surprising that health and safety issues should arise in the petroleum industry. Many refineries use more than 1,600 chemicals as additives or create them as by-products. As Barry Commoner has noted, "At each successive stage [of production], the industry produces more numerous, more varied, and more dangerous substances."[4] Most refineries operate twenty-four hours a day: it is cheaper to run the system until it breaks down than to stop production for maintenance or repairs. Recent studies have shown higher than expected death rates from brain cancer among refinery workers and high cancer rates in communities with a concentration of oil plants.[5]

Soon after the strike began, a coalition of environmental and public interest organizations announced its support for the OCAW demands and the boycott. These endorsers included the Environmental Defense Fund, Friends of the Earth, the Natural Resources Defense Council, the Wilderness Society, the National Welfare Rights Organization, and several chapters of the Sierra Club. In an advertisement in the New York Times, twenty-five scientists and educators who were part of the coalition explained their support: "Workers have long served as unwitting guinea pigs, providing useful toxicological data which helped to protect the public. Many toxic agents disperse beyond the plant and pose public hazards. The success of the OCAW strike is critical both to labor and the public."[6] The coalition also paid for billboards to explain the issues to the public and sent speakers to meetings in communities near the affected refineries. Not all environmentalists supported the union's demands, however. Several hundred members of the Sierra Club, primarily in oil states, resigned because they thought the group had gone too far in its support of a labor union.

After a five-month strike, Shell finally agreed to negotiate. OCAW won the right to receive information on sickness and death rates, but lost its other demands. Although this was only a partial victory, the action set an important precedent for new coalitions. Frank Wallick, the Washington, D.C., representative for the United Auto Workers, observed that "the Shell strike was a tremendous catalyst to increased awareness and support among labor and environmental groups."[7] In some places—California's Bay Area, for example—the strike led to coalitions and networks among trade unionists and environmental activists that are still active in the 1980s; their primary concern continues to be the health consequences of actions taken by the petrochemical industry.

Workers and environmentalists need each other. The factory gates that separate them by day do not stop the hazardous materials that sicken workers inside a plant from contaminating the air, water, or food both workers and nonworkers need outside. The companies that fail to protect their workers against toxic chemicals, that move abroad at the drop of a tax incentive, and that fight any attempt by workers to gain greater control over their working conditions are the same companies that dump their wastes near drinking water, pollute the air with toxic chemicals, and spray pesticides near populated areas. The Reagan administration's open support for big industry, its appointment of industry officials to head the Environmental Protection Agency (EPA) and the Occupational Safety and Health Administration (OSHA), and its attacks on occupational and environmental regulations emphasize that the government fails to protect both workers and community residents.

Yet despite shared problems and opponents, the relationship between the labor and environmental movements has been a troubled one. The primary conflict has been over jobs. For example, after antinuclear demonstrators occupied the Seabrook nuclear power plant, three thousand construction workers marched through nearby Manchester, New Hampshire, chanting, "Nukes, nukes, nukes."[8] Similarly, Edwina Cosgriff of Bring Legal Action to Stop Tanks (BLAST) reported that "union goons from New Jersey" disrupted their meeting because they feared that closing the liquid natural gas (LNG) tanks would lead to a loss of jobs.[9] Jim Sheets, research director for the Laborers' International Union, which represents more than 500,000 construction workers in the United States and Canada, gave a common labor view of the environmental movement: "As far as I'm concerned, a lot of these so-called environmentalists are a bunch of bloody elitists who don't want the view from their kitchen window messed up."[10]

Nevertheless, the two movements have much to offer each other. Organized labor's greatest potential contribution to the environmental movement is its political muscle. More than twenty million men and women in the United States belong to trade unions. At the national level, labor's support has helped to pass major environmental regulations. In 1981, for example, unions like the United Steelworkers of America, the International Association of Machinists, and the OCAW successfully lobbied for the defeat of an industry-sponsored initiative—backed by President Reagan—that would have weakened the Clean Air Act.[11] (Unions from the building trades, however, supported the initiative.) At the local level, unions have lobbied for new legislation such as Philadelphia's right-to-know bill, contributed money to environmental groups such as Indiana's anti-nuclear Bailley Alliance and joined coalitions.

The labor movement's experience in electoral and legislative work, its ability to call in political debts, and its political education funds are powerful resources to bring into a community struggle. Unions that can mobilize their members for demonstrations and public hearings can help to maximize the impact of these events. On the second anniversary of the Three Mile Island accident, for instance, 15,000 people converged on Harrisburg to demonstrate their opposition to nuclear power.[12] Many of the marchers were members of the unions that had helped to organize the demonstration. William Winpisinger, president of the Machinists Union, gave the keynote speech.

Labor's power is perhaps clearest in direct confrontations with corporations. Through a strike, workers can shut an industry down. The OCAW strike at Shell, and the health and safety concessions granted by the other major oil companies prior to that strike, illustrate a union's ability to win improvements in environmental conditions. That this has not happened more often is in part a result of labor's goals in the past three decades—primarily to fight for higher wages and job security—and in part because strikes on health and safety issues challenge management's control of the production process and are therefore difficult to win. Nevertheless, if community environmental groups can work with organized labor and thereby gain the ability to influence corporate decision making, they will be in a far stronger position than they are now.

Workers and their union can also contribute information. The harmful effects of exposure to substances such as asbestos, radiation, or polyvinyl chloride (PVC) were first discovered in the workplace. Workers often have detailed knowledge of toxic substances and their methods of disposal. And unions' ongoing battles with management give them experience and skills that can help community groups to learn where power lies and at what point public pressure can be most effective.

Environmentalists, in turn, have some quite different resources to offer the labor movement. One is their credibility and moral legitimacy. Some segments of the U.S. public see unions as organizations that express only the narrow self-interest of their members, an image reinforced by corporate propaganda. In the Shell strike, for example, management warned the environmental groups not to be "duped" by OCAW, which, Shell charged, was using health and safety issues to gain support for its real goal, the right to review pension plans.[13] Because some union leaders do give economic demands higher priority than health and safety, the appearance of narrow self-interest is perpetuated. The ironic result is that when workers do make working conditions a central concern, the public is often unwilling to support them.

Environmental groups can help labor to break down the myth of selfishness by emphasizing that occupational health problems eventually become community health problems. As Barry Commoner told a convention of members of the United Electrical Workers, "The environmental crisis in this country will not be solved unless you win your fight for decent working conditions, for health and safety measures in the shop."[14] By taking this message to their neighborhoods, environmental activists can help to build support for the labor movement.

Sometimes such support can make the difference between victory and defeat. In 1979, for example, workers at a uranium enrichment plant in Pikestown, Ohio, owned by the U.S. Department of Energy but operated by Goodyear Atomic, went on strike over both economic and health and safety issues.[15] After the strike had dragged on for seven months, a number of environmental and antinuclear groups organized a strike-support coalition that lobbied in Washington on behalf of the strikers. Political pressure finally led the Department of Energy to push Goodyear Atomic to accede to the workers' demands.

In times of high unemployment unions are on the defensive and their battles are often focused on preventing pay cuts or the loss of benefits. In such conditions a labor-environmentalist alliance becomes one way by which the labor movement can regain the initiative. When a company responds to demands for health and safety in the workplace with the threat to close the shop rather than make repairs, labor and environmental activists working together have a better chance of stopping this blackmail.

Labor unions can also benefit from the information environmentalists collect. A coalition of local fishermen and conservationists first called attention to the problems of polychlorinated biphenyls (PCBs) in the Hudson River. Their focus on the General Electric plant, which at one time was dumping thirty pounds of PCBs into the river daily, alerted workers in the plant to the hazard.[16] Local activists can also contribute energy, enthusiasm, and a range of tactics unfamiliar to unions. In situations where the workers are unable or unwilling to go on strike, community organizers can suggest other ways to build public awareness and support. Sometimes an environmental group can aid a union simply by opening another front: at the time a union local was on strike at a Long Island branch of the Hooker Chemical Company, a community group was piling garbage on the corporation's front lawn to protest the dumping of toxic chemicals.

Finally—and this is the result of all such coalitions—the environmental movement can help to break down the artificial distinction between community and workplace, which are generally perceived as two separate arenas. Work is what you do to earn a living; family, friends, and community are everything else. In fact, as we have seen, the two are inseparably linked. Toxic materials in the workplace escape into the general environment. Products made in the factory are consumed in the community. The control that corporations exercise in the political arena influences every aspect of our daily lives.

With its emphasis on economic gain, the labor movement in this country has tacitly accepted this artificial separation. As a result, some of its most powerful potential allies—women, blacks, ethnic groups, and consumers—have been unwilling to support labor demands that they could not see as being in their own interests. The environmental movement can help labor move beyond its obvious constituency by demonstrating the common interests, goals, and opponents that workers and community residents share.

But only 20 percent of U.S. workers are in trade unions and workers without unions have few protections against health and safety hazards, low wages, and poor working conditions. Coalitions of community environmental and labor organizations can sometimes begin the difficult task of reaching these workers. The Electronics Committee on Safety and Health in the California Bay Area's Silicon Valley, for example, united residents concerned about groundwater contamination originating from the electronics plants and workers exposed to toxic solvents. Similarly, in New

York City, the Washington Heights Health Action Project, concerned about the fire hazards posed by sweatshops located in residential buildings, convinced the International Ladies Garment Workers Union to give workers in these shops information on health and safety hazards and on how to join the union. Both these campaigns, however, faced immense obstacles in trying to organize undocumented workers who fear being deported.

Environmentalists must, however, avoid creating situations in which workers have to choose between jobs and health. If union members believe that environmentalists are willing to sacrifice jobs for a healthier community, they are likely to reject the swap. George McDevitt, a regional vice-president of the International Chemical Workers' Union and an ardent supporter of both labor and environmental causes, tells a story that illustrates this dilemma.[17] A community group in Brooklyn, New York, was fighting a factory that was polluting the air. A member of the group approached McDevitt for help in reaching the workers in the factory but, as McDevitt reported it, "They want to put the plant out of business and nothing less. How can I help them? Can I go to that plant and say to the people working there, 'If you sign up with me, I'll have this plant the hell out of here'?"[18]

Corporate leaders will exploit this division mercilessly. As the manager for special projects of the Public Service Company of New Hampshire, the sponsor of the Seabrook nuclear power plant, advised his colleagues: "Forget the facts once in a while. Counter the activists not with facts but with closed factory gates, empty schools, cold and dark houses and sad children."[19]

Some of the problems that have divided labor and environmentalists are easier to remedy. Community groups may be unfamiliar with the internal politics and channels of communication within trade unions. For instance, an antinuclear group that invited union members to its teach-in by mailing leaflets to all the unions in their area got not a labor turnout but a small pro-nuke demonstration by construction workers.[20] More careful homework and personal approaches to sympathetic leaders would have been more effective. In another case, a conference on the dangers of toxic chemical fires sponsored by environmental, community, and labor groups in New York City had to be canceled at the last minute because the planning committee had scheduled the meeting during contract negotiations for the firefighters' union.

Most labor unions are familiar with a well-defined, hierarchical decision-making process, so to be confronted by a coalition of environmental groups, each with a different viewpoint, can be frustrating. Gail Danneker, a member of Environmentalists for Full Employment—an organization that seeks to bridge the gap between the two groups—notes that "labor works by giving a little here, a little there. But you find some compromises that the Sierra Club will go along with, only to have Friends of the Earth turn around and sue the hell out of you."[21]

Environmental activists have their own problems with unions. They accuse some of being bureaucratic, undemocratic, even unprincipled. They criticize union leaders for accepting industry propaganda on the costs of environmental regulations and charge that unions are sometimes unwilling to join coalitions, even when it would be in their own interest to do so. Some of the labor groups fighting for right-to-know laws, for example, have refused to add provisions for a community's right to receive information because they believe that expanding the fight might complicate the issue, or reduce the chances for victory. As a result of these perceptions, environmental activists have sometimes been unwilling to work with labor groups. They fear that the alliance might force them to dilute their goals and lose sight of their vision.

Fortunately, there is a rich experience of labor-environmental cooperation. But to proceed successfully, activists from the two movements need to acknowledge their differences openly. Unions have as their primary goal protecting the interests of their members and of the organization itself. Local environmental groups usually want to remove a threat that endangers their community. While these goals may be compatible in the long run, in the short run they often conflict. Neither side gains from papering over the disagreement and open dialogue—even debate—can help to make clear the prospects for cooperation. Before such dialogue can begin, however, environmental activists need to learn about the labor movement. What issues concern local union members? What is their relationship to the national union? How many workers in the area are not part of a union? Without answers to these questions, a community group will find it difficult to know whom to approach.

The labor movement in this country is not monolithic, and unions are deeply divided on environmental issues: industrial unions tend to be more supportive of environmental protection, while craft unions are often against it. In part this difference reflects the specific hazards each group faces; in part it reflects their assessment of the cost in jobs of stricter regulation. Obviously, environmental groups looking for labor support should first approach unions known to be sympathetic. Local occupational safety and health committees, organizations of union activists, and health and safety professionals can often provide guidance as to where to start.

Sometimes the workers directly affected by a hazard are those least likely to support a community group because of their fears about jobs. In that case, it may be wise to approach another union first. If, for example, employees of a chemical plant rebuff an overture, community organizers might approach the firefighters union that would have to respond to a fire at the factory. The firefighters may be more effective in reaching chemical workers than the community residents were. This tactic can backfire, however, if it leads to major battles between unions.

Leaders and members within one union may have different priorities. Some surveys indicate that rank-and-file members are more likely than their leaders to value health and safety—given that it is the workers who encounter the hazards, this is not surprising. But community groups should first approach the union leadership (at the local or regional level); if they receive a negative response, they can then go to the rank and file. George McDevitt offers the following advice:

> If you go to a local president, you may get nothing because he doesn't understand. He may even be his own worst enemy. You must complete the whole thing right up to the top man in that particular union, and then you must document it. And if you don't receive a response then, go do whatever you want. . . . The Constitution allows us to distribute literature, so why not pass out flyers at the factory gate? Make it truthful, that you talked to these people in the union and that you're very concerned about the plant. And that if community folks are getting contaminated, so are workers.[22]

In some situations, workers will offer support outside the union context. At Love Canal, for example, Hooker employees regularly provided the Love Canal Homeowners' Association (LCHA) with behind-the-scenes information and assistance.

Labor and environmental groups find it easier to work together if the issue is defined so that both parties benefit from a victory. In

Philadelphia's right-to-know campaign, for example, both unions and community residents supported the new bill because neither had to sacrifice principles or the needs of their organizations in order to participate. Similarly, a coalition of community groups and OCAW Local 1.5 in northern California sponsored a Community Speak-out on Cancer: the high concentration of petrochemical industry in the area (Shell, Standard Oil, Dow Chemical, Ortho, and Chevron all have plants there) and its high cancer death rates made the topic of concern to both workers and residents.[23]

The effects of toxic chemical exposures on children and on future generations is another issue that links labor and community groups. Tony Mazzocchi of OCAW put it this way:

> The whole question of what's going to happen to our children is an issue that can easily be developed into a coalition base. Everybody is concerned about children. And what better place to start a dialogue. The real fight is over the rights of our kids to be born healthy and survive. That's an issue that emanates from the workplace and can incorporate broad sections of the population. Most workers feel they're working at lousy jobs because they want to make it better for their kids. If they doom their kids by virtue of having to work at these jobs, the indignation level is going to rise appreciably. I do believe that the 1980s are going to be a time for genetic confrontation.[24]

Another way to create alliances with labor is to look for ways that environmentalists can demonstrate a concern for unions' needs. A few examples illustrate this strategy. Members of the Clamshell Alliance set up a picket line at the Pilgrim nuclear power plant in Plymouth, New Hampshire, to protest the workers' exposure to high levels of radiation. More than a third of the six hundred workers at the facility respected the picketline.[25] In other actions, the Clamshell Alliance helped to publicize the national boycott of J. P. Stevens, a clothing manufacturer that refused to recognize its workers' right to organize, and raised money for the United Mine Workers during their strike. In another situation, the Harlem mothers who forced the New York City Board of Education to repair flaking asbestos in their children's school insisted that any repair procedure must protect the workers from exposure. And a Safe Energy Alliance chapter in New Jersey joined with local unions to organize for Solidarity Day in 1981, a national protest against Reagan's labor and social policies.

As a result of such activities, union members come to trust environmentalists. In the previously described strike at the Ohio uranium enrichment plant, Senator John Glenn tried to split the newly formed coalition between environmental groups and the union representing the employees at the facility by warning the workers not to trust their antinuclear supporters. But the strikers jeered his speech and local president Denny Bloomfield responded to Glenn by saying, "Listen, these folks [the environmentalists] have at least responded to our problem more effectively than anyone else in Washington. The fact that we have political differences doesn't matter. So don't tell us who our friends are."[26]

The developing trust between labor and environmental activists is more than mutual back-scratching. By working together, by supporting each others' struggles, and by debating when necessary, the two movements develop a common experience out of which can emerge a common political strategy for achieving mutual goals.

The coalitions described here are the first steps toward an alliance between the labor movement and environmentalists, but such an alliance will remain peripheral unless the environmental movement tackles the issue of jobs head on. In some cases, the arguments are relatively straightforward. When environmentally unsound practices lead to the elimination of jobs, then coalitions can demand an end to such practices. For example, single-crop agriculture permits mechanization that reduces employment while herbicides eliminate jobs in brush clearing. Similarly, the antinuclear movement has presented convincing evidence that investing in solar energy would create more jobs than investing the same amount in nuclear power. In yet another situation, opponents of the Westway highway project in Manhattan charged that not only would the proposed roadway increase air pollution, it would also create far fewer long-term jobs than an alternative plan to invest the money in mass transit improvement. Peace groups have also challenged the argument that military spending creates jobs, a challenge supported by economic research.[7] A resident living near the Seneca Army Base, a major storage site for nuclear weapons in upstate New York, claimed that the "base is contradictory to what is needed in this area. We need productive jobs here and the base doesn't supply them."[28]

Nevertheless, future or theoretical jobs do not pay the rent or food bills. Environmentalists who are serious about winning labor support need to demand retraining or guaranteed continued employment. "I've never met a worker who wanted to work in a polluted plant or live in a polluted environment," said OCAW's Tony Mazzocchi. He continued,

> People have to start understanding that workers bear the costs of environmental illness as well as the economic costs of shutdowns. If we intend to stop a project from being built, or from continuing to operate, the first demand that has to be laid down is that all workers are paid full pay for life. And I can tell you, you'd better stand out of the way as those workers trample out of that plant.[29]

Finally, and perhaps most importantly, ongoing cooperation between the labor and environmental movements requires both parties to challenge the capitalist belief that continued corporate growth is the only solution to the nation's problems. As long as people are asked to choose between employment and health, industry will be able to divide workers from others concerned with environmental issues. The real questions both movements need to ask are:

- Who benefits from the growth of corporations? Who loses?
- What is the ultimate purpose of growth and what are its consequences?
- Who decides where social resources are to be invested?

When workers and environmentalists can together articulate a vision of a society that meets human needs without destroying the life-sustaining environment, they will have taken a giant step forward.

People of Color*

Case History: Harlem, New York City

Public School 208 is located in the middle of Harlem, a black New York City neighborhood.[30] *When Helene Brathwaite took her child to school one day in 1978, she noticed a white powder flaking off the ceilings in the corridors. She had recently heard a*

---

*The term "people of color" will be used to refer to those of black, Hispanic, Asian, and Native American descent because the more common "minorities" sometimes serves to minimize the importance and power of these groups.

radio report on the asbestos problem in schools, so she became concerned. As the newly elected president of P.S. 208's Parents' Association, she thought she ought to investigate. After persistent inquiries, she found her fears confirmed: it was asbestos that was flaking off the ceiling. A work order had been sent in calling for repairs, but nothing had yet been done. The school authorities were not especially worried. They would correct the situation, they said, but it posed no imminent hazard to the children.

Brathwaite and other officers of the Parents' Association were not reassured. They had read that asbestos was one of the most potent cancer-causing substances known, and there were reports of children getting asbestos-related cancer twenty years after a brief exposure to dust brought home on their fathers' work clothes. Parents thought their children faced a lifetime of fear.

Using its own funds, the Parents' Association hired a private laboratory to conduct tests in the school. They also contacted researchers at Mt. Sinai Medical Center and the New York Public Interest Group, both of which had previously been involved in asbestos issues. These experts formed an advisory group that aided the parents in their negotiations with school officials.

When the investigators reported high levels of asbestos, the Parents' Association decided to act. At a community meeting called to present these findings, parents decided that the school must be closed until repairs could be made. But they insisted that their children's education not be disrupted: entire classes, with their teachers' must be bused to neighboring schools. The parents also demanded a voice in determining the procedures to be used to correct the hazard. Not only would they not accept a band-aid repair job, but they suggested specific ways to protect the workers making the repairs. Brathwaite and her co-workers also organized a telephone hotline to answer questions and dispel rumors. Regular media coverage and some support from local elected officials helped keep up the pressure on the school board, but a boycott of the school was needed to force official action. The school was closed for seven months and more than 850 children were transferred to seven neighborhood schools while repairs were made.[31] When the school reopened, most parents believed that it was relatively safe.

Newspaper and television reports of the parents' boycott helped to make asbestos a major political issue throughout the area. Senator Jacob Javits held hearings on the problem at P.S. 208. In 1979, in response to pressure from parents' and public health groups around the country, Congress passed the federal Asbestos School Hazard Detection and Control Act, which provided some funds for the detection of asbestos in individual schools.

But the Parents' Association had failures as well as successes. A plan for ongoing testing of the exposed children never materialized. This lack of baseline data will make it difficult to assess whether the asbestos is in fact the cause of subsequent health problems. And although the parents developed a model for ridding schools of asbestos hazards, few other schools have been repaired as extensively—despite a 1977 survey by the New York City Board of Education that showed that more than four hundred schools had asbestos problems.[32]

Why were Harlem parents relatively successful in getting their children's school fixed, where others had failed? In the late 1960s education, and particularly parent's right to participate in making decisions about their children's education, became a central political issue in Harlem, then New York City's most important black neighborhood. Suspicion and mistrust of the school system were pervasive and this made it easier to mobilize parents around a school problem. Not only was the school system not teaching the children, but it was also exposing them to a toxic substance; thus the issue was one of survival. According to Brathwaite, many parents felt that the reluctance of the authorities to clean up the asbestos promptly was simply one more example of the systematic oppression of blacks.

The support of white environmental groups and of noted scientists made it easier to convince the school officials to act. The technical expertise that the Parents' Association acquired from their consultants and their ability to communicate the problem to the media put the board on the defensive. Most important, however, the echoes of the civil rights movement gave the parents a moral and political legitimacy that the school board could not ignore.

There is a pervasive belief that blacks and other people of color are not interested in environmental issues. Yet the campaign to rid P.S. 208 of asbestos, the struggles of Native Americans to fight water contamination at the Pine Ridge Reservation, and the protests of black groups against the dumping of PCBs in Warren County, North Carolina, all show this to be a misconception. Moreover, nonwhites' support for environmental causes seem to be growing. National surveys show that the proportion of black Americans who believe that too little is being done to protect the environment increased from 33 percent in 1969 to 58 percent in 1976—while only 54 percent of whites thought this to be the case in 1976.[33]

People of color have special reason to worry about the environment because collectively they carry a disproportionate burden of the costs of industrial and agricultural pollution. Their jobs, access to health care, nutritional status, the conditions of their communities and their political and economic position within the country, all put them at greater risk of getting sick or dying from exposure to hazardous materials.

People of color work in the most hazardous jobs. Department of Labor statistics show that although blacks make up less than 11 percent of all workers in the private sector, they comprise nearly 16 percent of those in high-risk jobs.[34] Studies of rubber, steel, laundry, and shipbuilding workers show blacks to be in more dangerous job categories than whites and to have higher death rates.[35] Between 1950 and 1975, the age-adjusted cancer death rate increased 3 percent for whites but 20 percent for nonwhites.[36] (Age adjustment is a statistical procedure used to compare two populations after removing the effect of age differences.) The largest increase was for nonwhite males, the group most likely to work in high-hazard industries. As a result of their more dangerous working conditions, black workers have a 37 percent greater risk than whites of suffering from an occupational illness or injury and a 20 percent greater chance of dying from a job-related disorder.[37]

Other ethnic groups face their own specific hazards. Navajo Indians' frequent employment as uranium miners and their exposure to uranium tailings in their communities have contributed to an increasing lung cancer death rate for this group. Similarly, 49 percent of the approximately five million people engaged in migrant or seasonal farm work are Hispanic, and an additional 6 percent are black or from some other nonwhite population.[38] Agricultural work has the third highest accident rate among industries in this country. Farmworkers also face low wages, inadequate housing, poor sanitation, and continual exposure to pesticides. Young children come to the fields so their parents can watch them; older children are required to work themselves.

The widespread spraying of pesticides has affected blacks, Hispanics, and Native Americans more severely than whites since these groups more often reside in poor agricultural communities. In 1973–1974, scientists from the Environmental Protection Agency's (EPA) Ecological Monitoring Branch studied insecticide residues in human fat tissues from around the country.[39] Samples from blacks contained almost twice as much DDT (now banned) as did samples from whites. Another study found that only 45 percent of white newborns but 84 percent of black newborns demonstrated evidence of recent DDT exposure in the rural Mississippi Delta.[40]

Asian-Americans and Hispanics—mostly women—are disproportionately represented in the garment industry. Conditions in the sweatshops of New York and Los Angeles—thick dust, poor ventilation, chemical solvents, blocked fire exits—resemble scenes from another era. The problems of sweatshop workers are often compounded by the fact that many are illegal immigrants, which makes them reluctant to approach unions or government agencies to complain about their exploitation or the unsafe conditions.[41]

Inner cities pose a variety of health hazards, and almost 60 percent of blacks live there—compared with less than 30 percent of whites. These areas have higher levels of air pollution from industry and motor vehicles, higher rates of lead poisoning (the rate for black children is six times that for whites), and higher levels of carbon monoxide.[42] Not surprisingly, these conditions lead to higher rates of death and disease. A study in Houston, for example, showed that populations living in areas with higher exposures to air and industry pollution experienced higher mortality rates from lung cancer, other respiratory disease, and heart ailments.[43]

People of color are also less likely to receive the same level of government protective services than are whites. Garbage pickups, the enforcement of fire and air pollution regulations, and housing code inspections are more frequent in middle- and upper-income communities than in poor ones.

All the factors that damage the health and well-being of people of color also affect other low-income and working-class communities. As the case histories in this book demonstrate, toxic chemicals can and do poison anyone. But blacks, Hispanics, Asian-Americans, and Native Americans also carry the burden of class oppression and racism. Their multiple vulnerabilities have made environmental issues a question of survival—a problem that people of color have faced before. Their long history of struggle and their experience in fighting for their rights make these groups invaluable allies for the environmental movement.

People of Color's Contribution to the Environmental Movement

When environmental activists look for partners in their campaigns against exposure to toxic chemicals or other hazards, they ask several questions about their potential allies. How strong are they? What experience or skills do they have to offer? What is their stake in the status quo? How can they help to increase the movement's political power?

How hard people are willing to fight for change depends in part on their stake in the existing system. Those who benefit most from the current distribution of wealth and resources are less likely to engage in efforts that might have the long-term effect of challenging their own privilege. By almost any criteria—income, education, health, political power—people of color in the United States have received the fewest rewards from the country's wealth. As a result, no people would gain more from a society that was more just, more humane, and healthier. It is no accident, then, that a review of the history of the United States shows that people of color have often been in the forefront of struggles for change. For centuries, Native Americans have resisted attempts to destroy their land, culture, and people. Blacks fought first against slavery and then its legacy of oppression and racism. In the nineteenth century, Chinese-Americans organized against the brutal conditions facing workers building this country's railways. And in the twentieth, Chicanos led the drive to unionize farm workers.

In numbers alone, people of color are a potentially significant political force. According to the 1980 census, blacks constitute 11.7 percent of the population, Hispanics (black and white) 6.4 percent, Asian-Americans 1.5 percent, Native Americans .6 percent, and "others" 3 percent; the remaining 76.8 percent is non-Hispanic white. Thus more than 20 percent of the population—about 34 million people—are people of color. But the numbers tell only part of the story. In part because of their exclusion from the political system, people of color have been forced to develop their own strategies for change. And the lessons from this history of struggle can help environmental organizers to keep their movement strong and growing.

In those situations where people of color have taken on toxic chemical issues, they have firmly linked this problem with broader questions of human rights, self-determination, and their survival as a people. In Warren County, North Carolina, as we saw, black churches and civil rights organizations helped to convert a local struggle against PCB dumping into a national issue. The tactics used by the dump's opponents—mass arrests following nonviolent civil disobedience—further emphasized the connection between the civil rights and environmental movements.

In their battles against energy development, Native Americans in South Dakota, New Mexico, and elsewhere have demonstrated the relationship between environmental health and survival. Madonna Thunder Hawk, co-founder of Women of All Red Nations (WARN), explained her group's opposition to uranium mining:

> This whole uranium thing is very dangerous to our survival as a people. We have no other gene pool anywhere else in the world and this radiation and chemical contamination with all the herbicides and pesticides they're spraying—we know their effects on genes. We know our very survival is at stake. We can't for many years afford the spontaneous abortions, the sterilization abuse, the contaminated drinking water, the acid rain, the cancer rates, and that type of thing. And we are charging the energy corporations and the U.S. government with genocide. We are fighting this genocide full strength. Our greatest asset is the power and drive of our people.[44]

White environmental activists can benefit from this broader definition of the problem in two ways. First, insisting that the real question is survival makes the seriousness of the issue clear.

Middle-class Americans are often reluctant to believe that corporations or their government could deliberately engage in activities that might sicken or kill people, but the historical experience of nonwhites has led them to believe otherwise. They are thus more willing to engage in militant activities in order to bring about change.

Second, it makes more difficult industry efforts to separate environmentalists from other sectors of the population by charging them with elitism. Blacks and other minorities need jobs and growth, the corporations claim, and the environmental movement opposes them.

If opposition to hazardous industrial projects includes people of all colors, it is more difficult for corporations to accuse their foes of being motivated only by "selfish" concerns about wildlife or pristine nature. Insisting that hazardous projects threaten jobs rather than creating them further strengthens these coalitions. In short, what alliances with blacks, Hispanics, Asian-Americans, and Native Americans offer white organizers of environmental struggles is the opportunity to break out of the isolation that the movement has sometimes experienced, either at the hands of its corporate opponents or as a result of its own practices.

Issues Separating People of Color and Environmentalists

Although people of color and the environmental movement need each other, the fact remains that they have had a difficult relationship. The following examples illustrate the range of problems:

- During Earth Week at San Jose State College in 1970, a student environmental group bought a new $2,500 car and buried it to demonstrate their opposition to a consumer society. Black students picketed the event, arguing that the money could have been better spent on improving conditions in the inner cities.[45]
- In 1972 the membership of the Sierra Club was polled on the question of whether to increase involvement with the urban poor and minorities. The proposal was rejected three to one.[46]
- In 1977 the National Association for the Advancement of Col-

ored People voted to endorse nuclear power as a necessary condition for economic growth.[47]

- During the crisis at Love Canal, tensions developed between the mostly white homeowners and the mostly black renters in a nearby housing project. When leaders of the Love Canal Homeowners' Association (LCHA) visited a meeting of the renters' group to ask them to join their effort, their entreaties were seen as "self-serving attempts to increase the size and power of the Homeowners' Association."[48] The distrust of the LCHA was fueled when it refused the renters' demand for equal representation on its board of directors. As a result, LCHA leaders were asked not to return. The black renters also thought that the whites were getting more attention and resources from the state.

It is hardly surprising that the racial conflict that infects this country should be found within the environmental movement as well. By understanding its causes, activists can begin the search for solutions. Predictably, people of color are not interested in joining organizations that do not take their needs seriously. Until recently the national environmental groups have had almost no nonwhite members, so, like some labor leaders, people of color have perceived the movement as being willing to sacrifice their concerns for an abstract dedication to wildlife, clean water, or national parks.

Indeed some environmentalists believe that the movement *should* define its goals narrowly and specifically. "You can't be saving national parks and improving minority housing at the same time," said one. "We have to make our political force as specific as possible."[49] But the decision to opt for limited objectives reflects an acceptance of the broader status quo and ignores both the common causes of environmental degradation and other social problems and the potential for more powerful alliances.

At the local level, people of color often think that whites have an easier time getting the attention of government officials. In an interview, Margaret Williams, a black activist from Memphis, Tennessee, noted that "state officials try to divide people. They call meetings and invite only some people. In Frayser, they evacuated some white families, but in Hollywood, which is black, they didn't evacuate anyone."[50] Such preferential treatment further divides groups that may already have an uneasy relationship and if white groups do not address these concerns, they are unlikely to win over organizations representing people of color.

What steps can environmental groups take to increase the likelihood of successful coalitions or alliances with people of color?

First, white environmental activists need to reach out to these communities: if people do not know about a hazard they cannot fight it. This may mean developing educational materials in Spanish, Chinese, or French Creole. It may mean speaking to representatives of community organizations. The Washington Heights Health Action Project, described in chapter 3, discussed the issue with the many political and social clubs organized by the area's immigrants from the Dominican Republic.

Second, the issues must be defined in a way that invites participation. This means emphasizing how hazardous materials affect *all* people. Sister Jacinta Fernandes of the Coalition for a United Elizabeth explained how this was done in Elizabeth:

> The Chemical Control fire really affected a variety of groups—we have an area down there that's predominantly black, an area that's predominantly Polish-Lithuanian, Portugese, we have Hispanics, so there are diverse groups that have been affected. . . . I think because people experienced it, they were able to relate to one another on the issue. It was definitely a unified thing. The ones that were seen as the enemy were the government officials and the people from Chemical Control.[51]

It also means addressing head-on the issues likely to concern people of color: specifically, jobs, economic development, and fears of genocide. When hazardous industrial projects are shown to threaten jobs as well as health, when alternative, safer patterns of development are proposed, when the links between economic and environmental survival are clearly made, then coalitions between environmental and nonwhite groups proceed more smoothly.

At a 1981 New York City conference for environmental activists, a workshop on environmental hazards in black and Hispanic communities recommended that major organizing for environmental action in Third World communities must be done by those who live in these communities. Because black and Hispanic organizers are more familiar with the culture and values of their communities, they are more likely to be successful in educating and mobilizing their own people. Just as important, since many people of color see self-determination as a goal of their struggles, and think that whites have been making decisions for them for too long, they have decided to create all-black or all-Hispanic or all-Native American organizations to achieve their goals. This decision does not necessarily preclude coalitions with other groups, and environmental organizers need to respect this choice and explore ways to work together.

Finally, if environmental organizations hope to build ongoing alliances with people of color, they have to commit themselves to the fight against racism, prejudice, and discrimination. This fight takes place in many arenas. Activists must become aware of subtle and not-so-subtle prejudices within their own organizations. For example, an integrated coalition that was fighting toxic chemicals elected an all-white slate of officers. Intended or not, this appeared to the blacks to be a statement that they were neither welcome nor equal partners.

There are strategic as well as moral reasons for environmental activists to combat racism in their organizations. Lois Gibbs described how she was invited to speak to a group opposing a toxic waste facility. When she entered the meeting room, blacks were sitting on one side, whites on the other. "By your seating arrangement," she said to the group, "you have told your opponents how to divide you."[52]

On a larger scale, ongoing alliances require mutual trust. If environmental groups, through their actions, demonstrate that they are opposed to any policy that imposes an unfair burden of risk on people of color, then a partnership between the two forces can proceed. But too often a middle-class community will oppose a hazardous project in their area, only to accept it if it is put in a low-income neighborhood.

Historically, racism has been the primary tool used to prevent poor and working-class whites from joining with people of color to fight for a more just society. Those in power have handed out a few crumbs to poor whites telling them, "You could be worse off. You could be down there with the blacks." An environmental movement that can successfully challenge polluting corporations must learn to anticipate and reject these attempts to divide its potential supporters.

Feminists

Case History 1: Women Opposed to Nuclear Technology, Huntington, Long Island

Women Opposed to Nuclear Technology (WONT), a group in Long Island's Suffolk County, was created in April 1979, soon after the Three Mile Island accident.[53] Concerned about their proximity to the planned Shoreham nuclear plant, a group of women decided to organize in their own communities. According to Charlotte Koons, a teacher in the group, it was "women who wished to work with women" who joined. WONT participated in antinuclear demonstrations and marches, made presentations at Town Board meetings and at other public hearings, circulated petitions, and initiated telephone and letter-writing campaigns to elected officials.

Like the women's movement from which it sprouted, WONT sought to achieve its aims through education and consciousness raising and made a special effort to reach women: its members spoke at meetings of local chapters of the American Association of University Women and the National Organization for Women and sponsored events with the Women's Health Alliance, a feminist health organization, and the Women's International League for Peace and Freedom. A theater piece entitled Spin the Web Sisters—an "original ecological antinuclear production"—was an integral part of the community education campaign. In its presentation, WONT focused on the effects of radiation on human reproduction and also emphasized the connection between nuclear power and nuclear weapons. In 1980 and 1981, WONT members helped to organize the Women's Pentagon Action, national demonstrations against U.S. militarism and weapons policies.

WONT was particularly interested in developing the organizational skills of its members. Thus one woman was responsible for coordinating the theater production, another for planning testimony at public hearings. Leadership was rotated so that every individual became a key member of the group. With their new leadership experience, WONT members became organizers in other feminist and environmental efforts.

As of early 1984, the Shoreham plant had still not become operational due to widespread opposition by state and Suffolk

County officials and residents. Some officials predicted that the plant would be abandoned. WONT's unique contribution to the anti-Shoreham coalition was to wed feminist concerns and organizational forms with local antinuclear sentiment. It brought new people into the campaign and, by linking nuclear power with nuclear weapons, broadened the perspective of the coalition as a whole.

Case History 2: Women's Pentagon Action, 1981, Washington, D.C.

On November 15 and 16, 1981, 3,500 women came to Washington, D.C.,[54] to demonstrate at the Pentagon because, they said, "We fear for our lives, we fear for the life of this planet, our Earth, and the life of the children who are our human future. Life on the precipice is intolerable [and so] we will not allow these violent games to continue."[55] Bringing together women organizers from around the country, the Women's Pentagon Action illustrates the potential for coalitions that include feminists, peace activists, and environmentalists. It also shows how the feminist experience in creating democratic and participatory organizations can contribute important lessons to the environmental movement.

Planners of the action did not want to stage another mass rally because too often, it seemed to them, such events have little impact on either the demonstrators or those in power. Therefore, as Donna Warnock, an organizer of the Pentagon Action, explained, "We wanted it to be a totally participatory event which reflected each woman's unique commitment and contribution."[56] Several activities were planned to achieve these goals. The first day was confined to preparation, and each woman participated in planning some aspect of the next day's action. To help break down the anonymity that goes with large groups, each woman wore a name tag with her own first name and two last names—Silkwood (after Karen Silkwood, a union organizer in a nuclear processing plant who had died under mysterious circumstances) and Ward (after Yolanda Ward, a black urban activist murdered in 1980). The common surnames emphasized the bonds between all women who fight against injustice.

Each demonstrator also made a cardboard tombstone for

women who were victims of oppression in their own communities. Some commemorated victims of rape or battering, others of radiation or toxic chemicals. One participant, Rhoda Linton, described reactions to this "cemetery": "What happened was that people's personal experiences, when they were put out into the public arena side by side showed fantastic patterns of similarity. So again it was a way of showing the individual as part of the collective without losing her individuality."[57]

Since one goal was to help women express how they felt about the oppression of women and the warlike mentality of the United States as symbolized by the Pentagon, the action on the second day used puppets, songs, banners, and theater to move participants through four emotional "stages"—mourning, rage, empowerment, and defiance. Warnock explained how acknowledging these emotions became a political experience:

> Because we were dealing with emotions, we were getting at people's humanness. And once you start getting to their humanness, what it means is that the skills that need to be relied on are not the kind of political skills generally used nowadays. We were relying on our skills as friends, as sisters, as mothers, as daughters, as lovers on very personal levels. And on that level, women are experts. So, as organizers, we were able to treat each other in radically different ways than people have come to expect from politicos. Just being there dealing with our emotions creates a feeling of power. It's a sensitivity that is crucial if we're going to turn things around because the reality is that change requires emotional growth that is sometimes painful.[58]

The movements against nuclear power and nuclear weapons have provided fertile meeting grounds for feminists and environmentalists. In the first part of this section we will explore the unique contributions that feminists have made to the environmental movement. Since many women who are not feminists have played a leading role in community struggles against pollution, we will also examine what feminists can learn from these organizers' successes in mobilizing other women. Finally, we will look at some of the obstacles blocking coalitions between the women's and the environmental movements.

The women's movement emerged out of the civil rights, student, and antiwar movements in the late 1960s and early 1970s. With intellectual and philosophical ties to the earlier suffragist movement, feminists insisted that women lacked the right to make decisions about reproduction, work, and personal relationships. They attributed this lack of freedom, and its resulting oppression, to patriarchy, which one feminist has defined as a "sexual system of power in which the male possesses superior power and economic privileges."[59] Although agreeing on the existence of oppression, feminists differed in their analysis of its *causes*. "Radical feminists" saw women's status as rooted in their biology, which often led them to call for a separatist women's movement. "Socialist feminists" attributed women's position to their economic class *and* their gender. Equality for women thus required a simultaneous attack on patriarchy and capitalism.

Both groups made important contributions to the analysis and practice of the environmental movement. For instance, one group, known as ecofeminists, argued that the oppression of women under patriarchy and the pillage of the natural environment are two aspects of the same phenomenon. The male power structure that sees women as instruments to be used for the economic, sexual, or psychological benefit of men also views natural resources as commodities to be exploited, then discarded. Thus, the struggle to liberate women cannot be separated from the effort to preserve the earth that sustains human life.[60] This perspective was an important advance for the environmental movement because it encouraged others to look beyond the superficial manifestations of the problem to its root causes, and because it connected the environmental dilemma to the question of who has power in this society and how they use it. Thus, by linking the destruction of the environment with the political and economic exploitation of women, feminists made a persuasive argument for a broad-based movement. This analysis is similar to the Native American contention that the colonial status of their people and the destruction of their lands are both based on a mentality that sees humans and natural resources as objects to be exploited.

Feminists have also made practical contributions to the environmental movement. WONT's theater piece and the Pentagon Action's "cemetery" illustrate the innovative forms of organizing, educating, and demonstrating that feminists have created. As described in chapter 5, community environmental groups have sometimes had difficulty creating organizations that can maintain their members' commitment. Feminist experience in group process, leadership development, and democratic decision making provides a valuable resource for environmental organizers.

The concern of the women's movement for the emotional impact of oppression also provides important lessons. Disasters such as the dioxin dumping at Times Beach, Missouri, the discovery of toxic chemicals at Love Canal, or the accident at Three Mile Island unloose strong emotions: anxiety, fear, outrage. Unless activists can help people understand these feelings, and convert them into energy for action, their prospects for organizing in these situations are bleak. The women's movement has developed powerful techniques for helping people transform their personal feelings into political commitment and action. The consciousness-raising group, the speak-out, street theater—all are methods that the environmental movement can borrow from feminists to help people convert their emotional reactions to toxic hazards into strategies for change.

Despite the influence of the women's movement on the environmental movement, most women involved in community organizing against toxic exposures do not define themselves as feminists. The women described in several case histories—Lois Gibbs at Love Canal, Helene Brathwaite in Harlem, Edwina Cosgriff in Staten Island, Carol Froelich in Rutherford—became leaders in their communities because they were mothers, activists, or church members. Yet in these leadership roles, women changed how they thought about themselves and their place in society. They learned new skills, gained self-confidence, and to a certain extent exercised political power.

Through confrontations with government agencies or corporations responsible for the pollution, these women activists developed a new understanding of the political system. As one Staten Island woman involved in BLAST put it, "Now I feel I can get involved and make a difference." Helene Brathwaite described what she had learned from her experiences: "I'm as competent now in the problem-solving area as they [the experts] are. Automatically, I expect flak, lies, placation, pacification, that people want me to go home and go to bed. But if a problem comes up again, I know exactly what to do."[61]

In sum, while environmental activism may not have led women

to define themselves as feminists, it has brought thousands of them into local political activity. It has strengthened their self-esteem, sharpened their analytic skills, and given them leadership roles.

What can feminists learn from this? Although the women's movement has had a profound—if often indirect—impact on society, it has had particular difficulty reaching working-class women and women of color. The extent of women's active involvement in community environmental struggles shows that women can be mobilized against threats to health, family, and community. Unlike the New Right, which has also tried to organize working-class women to fight for family and community, the community environmental movement defines its opponents not as blacks or feminists or people on welfare, but as those corporations that pollute and those government agencies that protect the polluters.

For feminists who seek to take their message to new constituencies, environmentalism offers both issues and methods that provide a bridge for broader coalitions. Issues such as chemical hazards to reproduction and the exclusion of women in their child-bearing years from jobs that pay well but require exposure to toxics can bring new people into reproductive-rights struggles. The use of organizations such as PTAs, church groups, and neighborhood associations; the successful use of mass media for community education; and the ability to force specific corporations to give in to demands are methods that feminists might borrow.

Problems Between Feminists and Environmentalists

We have seen that mutual benefits of cooperation have not always been sufficient to overcome conflicts between environmental activists and other groups. This is also true for feminists and environmentalists; real issues divide them and their solution requires both political analysis and the painstaking task of working out differences in practice.

A central issue for the women's movement in the 1970s and early 1980s has been reproductive freedom. Feminists believe that the right to decide whether and when to have children is a necessary condition for women's equal social, political, and economic participation in society. Reproductive freedom would require safe and legal abortion, contraceptive services, paid maternity leave, and high-quality, low-cost child care. In response to growing demands for these rights, a coalition of conservative forces launched a vigorous counteroffensive. The New Right, the Catholic Church, the Reagan administration, and some corporate leaders have formed an alliance—albeit an uneasy one—dedicated to restricting access to abortion and contraception and to keeping women in the home.

In the face of this attack, the women's movement has increasingly defined its friends and enemies by their willingness to support its stand on reproductive freedom. Environmental groups that want to work with feminists will have to address this issue. While not every local coalition against toxic waste will take a political stand on abortion, a coalition that includes active opponents of reproductive freedom—a Right to Life group, for instance—cannot expect the support of feminist organizations. For example, in 1981, labor, environmental, and women's groups organized the Coalition for the Reproductive Rights of Workers (CRROW). Its purpose was to advocate policies that would protect both men and women from workplace exposures that could damage their reproductive abilities. When certain members of the coalition wanted to invite Right to Life groups into the organization (they believed that any group that supported the coalition's overall goal should be encouraged to join), feminist members responded with outrage. To them, the inclusion of right-to-lifers would be like asking groups who oppose the existence of trade unions to join a labor coalition. Some labor leaders had trouble understanding this position; one said, "The perception that [the issue of reproductive hazards] is a women's issue is wrong. It's a workers' issue." Feminists replied that this distinction missed the point: toxic chemical hazards are both a workers' issue *and* a women's issue. Not only are a significant proportion of workers women, but companies have also used the potential danger these chemicals posed to a fetus as a rationale for excluding women from higher-paying jobs.[62] CRROW eventually disintegrated, in part because these conflicts were never resolved. Future organizers of collaborative efforts between labor, environmental, and women's groups will need to learn how to understand and respect their allies' principled positions, and how to resolve differences.

As we saw in chapter five, men and women members of environmental organizations are sometimes treated differently. Men are more likely to be spokesmen, decisionmakers, and leaders.

Women are more often found filing, licking envelopes, or typing. Not only does this deprive a group of the wisdom of its female members; it also sends a message to the community that it is a male organization. Thus, women's groups are less likely to form a coalition with such a group.

The solutions to these problems are similar to those recommended for facilitating coalitions with other groups. First, feminists and environmentalists need to engage in a dialogue on how the two movements can work together. In some situations, feminists will decide to organize within a broader coalition—as the case of WONT illustrates. In others, they will create autonomous organizations such as the Pentagon Action. In any case, frank discussions at both the local and national levels will contribute to the mutual understanding and trust that are the prerequisites for successful coalitions.

Second, environmental activists need to define issues in ways that attract feminists. Giving women the right to know what substances they are being exposed to in the workplace, community, and home; protecting them against environmental hazards to reproduction; and relieving women of the burden of having to take over responsibility for protecting their communities because of lax government regulations are all examples of issues that can bring the two movements together. Finally, environmentalists need to understand and respect the feminist agenda. Unless community groups support feminists' positions on reproductive rights and sex discrimination, for example, they cannot expect women's groups to provide organizational support for environmental campaigns.

Women hold up half the sky. By learning how to include feminists in their coalitions and by understanding how the oppression of women helps to maintain the status quo, the environmental movement will win a major ally in keeping that sky clean.

Peace Groups
Case History: Stop Project ELF

In the North Woods of Wisconsin and Upper Michigan, during the last decade, the U.S. Navy constructed a vast underground antenna system known as Project ELF.[63] Its purpose is to communicate with submarines, giving them the order to surface for firing instructions without attracting the attention of an enemy. Using extremely-low-frequency (ELF) electromagnetic radiation, the antennae send millions of watts of electrical energy deep into the northern bedrock, radiating an electric field that the submarines can detect. In 1982, Rear Admiral William D. Smith told the Wisconsin Natural Resources Board that Project ELF is "the most effective single thing that we can currently do to ensure the survivability of our submarines in a nuclear attack."[64] The U.S. Congress appropriated $49.8 million for ELF in fiscal year 1983 alone and previous expenses for the project had totalled $150 million.

But as the antenna cables snaked under the forests, a widespread opposition movement emerged. It began in the late 1960s and early 1970s, as residents learned of the navy plans to push through a large, expensive communications project. Known first by the code name "Sanguine," then as "Seafarer," these projects were stopped by public opposition from residents who feared their environmental consequences. Then, in 1969, a twenty-eight-mile antenna system was built in Wisconsin's Chequamegon National Forest. In 1978, the navy proposed another antenna for northern Michigan. It too was opposed by nearby residents, who organized a group called Stop Project ELF.

One task of the group was to collect all the information on ELF they could find. The available data were scanty, but the little the organizers could find disturbed them. Electric fields similar in intensity and frequency to those of Project ELF have been associated with poor driving performance and road accidents, suicide, nervous system impairment, and cardiovascular disease. The navy's research had suggested that ELF waves disrupted birds' migration patterns, altered growth in plants, led to changes in fat levels in human blood, and distorted the time perception of monkeys. Because most of the research had been funded by the navy or the utility companies, opponents of the project doubted that the information they had acquired told the whole story.

The group feared Project ELF's effects on humans and wildlife; in addition, construction of the project meant digging up areas of forest, disturbing conservationists and hunters. Further, the group had strong objections to ELF's military purposes, since it could trigger a first strike by alerting submarines for attack without attracting Soviet attention. In addition, the antennae made the North Woods an obvious target for an opponent's missiles.

Some of the navy's plans for ELF illustrate the bizarre thinking that characterizes U.S. military policy. In fiscal year 1982, for example, Congress voted to spend $5 million to develop a Mobile ELF, which would consist of vans carrying ELF antennae; the vehicles would be stored in tunnels in Michigan and Wisconsin. After a nuclear attack, they would venture out to lay a new ELF grid to reestablish contact with surviving submarines.

Some critics doubted that ELF would be effective and reliable even on its own terms. A 1979 General Accounting Office report concluded that "although GAO does not believe the ELF is needed, there is doubt that the system will work as planned even if it is needed."[65]

Finally, Stop Project ELF opposed the plan because citizens feared it would provide the basis for a much larger military system and for other dangerous projects. Jenny Speicher and John Stauber, coordinators of the group, argued, "Project ELF would be a foot-in-the-door for other government and corporate schemes for the north country: disposal of civilian and military radioactive waste, uranium mining; nuclear power on the shores of the Great Lakes; and nuclear missiles in Lake Superior."[66]

The diverse objections helped Stop Project ELF to recruit a broad spectrum of support. At the local level, the organization worked with labor unions, food-buying clubs, tourism business associations, veterans groups, vacation homeowners, and Native Americans. Local, regional, and national peace and environmental groups such as SANE, the Fellowship for Reconciliation, the Audubon and Sierra Clubs—among others—joined the campaign. By early 1980 the group had more than one thousand members.

Stop Project ELF also had its opponents. GTE Sylvania, a giant multinational corporation, was the prime contractor for the project. Its heavy reliance on government dollars (in 1979 the company won $227 million in government contracts for military goods and services) made GTE an ardent advocate of the project. The corporation helped to organize a small network of local citizens, mostly its own employees, who red-baited opponents, confronted them in the media, and distributed pro-ELF information.[67]

By 1982, however, public opposition to Project ELF was growing. Wisconsin voters passed a nuclear freeze resolution by a wide margin, while Anthony Earl, a liberal Democrat and an ardent freeze supporter, was elected governor of Wisconsin. In his campaign, Earl pledged to "eliminate Project ELF from the Defense Department's agenda once and for all."[68]

In September 1983, anti-ELF activists tore up and burned hundreds of surveyor stakes along a twenty-mile route in Michigan's Upper Peninsula, bringing the struggle against the navy's project to a new level of militancy.[69]

Stop Project ELF organizers believe they have "built a network of people that is growing in strength and ability to influence politicians." In the future, Stop Project ELF hopes to ally itself more closely with the national peace movement so as to cut off funding for the project at its source in Washington. But until deployment of ELF is stopped and existing facilities are dismantled, the group will continue its activities at the local and national levels.

Since the movement for a nuclear test ban in the late 1950s and early 1960s, peace activists have consistently raised the issue of the environmental consequences of military projects. Scientists began to educate the public about the health hazards of the aboveground testing of nuclear weapons. As a result, parents who objected to giving their children milk contaminated with cancer-causing Strontium 90, a by-product of nuclear testing, inundated their senators with angry letters. Eventually a mass movement forced the United States and the Soviet Union to sign the 1963 test ban treaty.

Other weapons also aroused opposition. In the 1960s public outrage at the continued use and testing of nerve gas forced the army to destroy gas stored at bases in Colorado and elsewhere. In the mid-1970s, Colorado peace and environmental activists joined to form the Rocky Flats Action Group, whose goal was to close down the Rocky Flats Nuclear Weapons Facility, located sixteen miles from Denver. The group's opposition was based both on the health hazards posed by the facility and its role in military policies that threatened the destruction of the environment and human civilization.

More recently, the growing debate about Agent Orange, a defoliant widely used by the United States in Vietnam, provides another example of the overlapping concerns of environmental and peace groups. Veterans organizations worried about the health effects of exposure to this substance have joined with environmentalists and, in some cases, Vietnamese scientists and health officials to demand more research on the effects of Agent Orange and services for its victims.

The movement against war and militarism has much to offer to the fight for a healthy environment. Fueled by Reagan's foreign and military policies, the peace movement has grown rapidly and the issue of war and peace is now firmly on the national political agenda. Obviously, military policies that increase the likelihood of nuclear war endanger everybody. But such policies also have a more immediate negative effect. The Reagan budget took money from social services, health care, and environmental protection and gave it the military and its corporate contractors. It is hardly accidental that budgets and staff for the EPA and OSHA plummeted as those for the Pentagon skyrocketed. In a very pragmatic sense, environmentalists gain from a strong peace movement that can use its power to divert tax money from destructive purposes to programs that meet human needs.

At the local level, the issues raised by peace activists can also help. Organizers against the ELF system in Wisconsin and Michigan, the Rocky Flats project in Colorado, and the planned MX missile sites in Montana and Wyoming have found that antimilitary sentiment added yet another reason to oppose these projects. President Reagan's proposal to use spent fuel from civilian nuclear power plants for military purposes promises similar possibilities for cooperation between antinuclear-power and peace groups.

Although the peace movement includes a variety of political perspectives reflected in a range of organizations, it is clearly a nationwide movement with national and international goals. This broader viewpoint can be a useful antidote to the localism of many grass-roots environmental organizations. By linking up with peace groups, environmental activists can expand their outlook. With this broader political understanding and with new contacts in statewide or national networks, activists will be better placed to win their environmental objectives as well.

Peace and environmental groups share enemies as well as goals. The same corporations that pollute the air, water, and land profit from bloated military budgets. Among the major producers of nuclear weapons are the Bendix Corporation, General Electric, Union Carbide, Monsanto, and DuPont, each of which has a long rap sheet of environmental violations. If groups opposed to these corporations' practices can find common themes or strategies, they will increase their ability to put pressure on corporate directors.

Peace activists can reap equally important benefits from their environmental colleagues. The movement against pollution has developed a number of techniques and strategies that antimilitary organizers have found helpful. For instance, the National Environmental Policy Act of 1970 gave supporters of a clean environment the right to make government and corporate sponsors of new projects file an environmental impact statement (EIS), a document that describes the effect of a proposed project on the local area. Several peace groups have used the EIS as a focus for debate. In Texas, for example, the Panhandle Environmental Awareness Committee forced the Department of Energy to issue an EIS on a planned expansion of Amarillo's Pantex facility, where nuclear warheads are assembled. The necessity of preparing an EIS stalled the project, and also sparked discussion on its perils and benefits.[70] Anti-MX missile groups have used the air force's EIS to demonstrate the ecological damage it will cause.

Environmentalists have found that threats to health can help mobilize people against a military project. A description of the effects of an accident, spill, or leak at a military facility can often spark widespread opposition. Carol Rothman, a staff person for the Rocky Flats project, explained that "many more people are more concerned about health effects than disarmament at first," and only later go on to consider the "global effects." Rothman also said, "I resent having my health and safety threatened for a national objective I don't agree with."[71] By linking day-to-day concerns with a vision of a more rational military and foreign policy, organizers lay a foundation for a strong movement that can survive over time.

The rapid growth of the peace movement in the last few years has meant that its supporters vary in their political sophistication and analysis. The nuclear freeze movement includes retired generals and prominent government officials as well as millions of ordinary citizens. On the one hand this diversity has contributed to its appeal. On the other hand, it has sometimes led to a lowest-common-denominator politics. Some critics of the freeze movement fear that its lack of analysis and leadership could lead to its rapid demise or co-optation. A member of northern California's antinuclear Abalone Alliance observed, "Sure, there are a lot of sincere and intelligent disarmers in the freeze campaign along with the cynical manipulators and professional bandwagon-jumpers. But I've seen a lot of worry about respectability and little cultivation of rebelliousness."[72]

Environmental activists, who have experience fighting corporations and government agencies, and who have networks of support within their community, may be able to give the freeze movement the neighborhood roots and the practical political direction it needs. In many communities, activists have already taken on this role; for example, some of the organizers of the victorious New Jersey 1982 nuclear freeze ballot referendum had first become involved in political activity in environmental struggles in Rutherford, Elizabeth, and Newark.

Healthy alliances and coalitions require conflict and debate as well as cooperation. Many of the issues that bring peace and environmental groups together also create controversy between them. For example, those who are opposed only to the *environmental* consequences of a weapons system might be willing to settle for a new location for the system, believing the weapon is necessary for national defense. Some peace activists, on the other hand, are opposed to *any* weapon that increases the likelihood of war. Even if such a weapon posed no threat to local residents, they would still be against it.

Some environmental activists argue against taking on military issues because it dilutes their other efforts. They fear that criticizing the U.S. military might alienate potential supporters or open them to charges of being unpatriotic. Advocates of linking the two issues reply that a nuclear war would make all other issues irrelevant. Anything that strengthens the peace movement, they say, improves the chances of the survival of the environment.

These differing perspectives can create problems. In South Carolina, for example, the Energy Research Foundation (ERF) opposed the restarting of a reactor at the Savannah River Plant, which produces most of the plutonium and tritium for the nation's nuclear weapons. The group wanted the plant to conduct an in-depth study of the reactor's environmental impact. However, the ERF did not want to risk achieving its limited goal by challenging South Carolina's commitment to military-related industries. As a result of this stance, it was able to win support for its position from unlikely sources. An organizer for the group explained, "I've taken this petition [to oppose further shipments of high-level nuclear waste into the state] into very conservative boardrooms and had every member sign it."[73] In this case, the environmentalists chose to solicit help from powerful business interests and other more conservative sectors of the population rather than from the peace movement. In the long run, such a choice does not help build an ongoing movement.

Similarly, some peace workers believe that their movement should focus only on disarmament or the nuclear freeze. The incorporation of other issues, they say, can confuse the public, dilute the message, and alienate potential supporters. Physicians for Social Responsibility, a national organization that educates people about the effects of nuclear war, has argued that in the face of the threat of nuclear extermination, all other topics shrink into insignificance. Environmentalists who want to raise their issues together with those of the peace movement may have to persuade antimilitary organizers to broaden their definition of the problem.

No one concerned about the fate of the earth can ignore the issue of war and peace. The exploding U.S. "defense" budget; U.S. military activity in Central America, the Caribbean, and the Middle East; and the deployment of cruise and Pershing II missiles in Europe make military policy a central political issue for the 1980s. The peace movement has mobilized millions of people in the United States and Europe. The immediate health hazards that many military projects pose as well as the diversion of resources from domestic programs to the armed forces provide environmental activists with obvious entry points for peace work. The growing alliances between these two movements promise new opportunities to achieve their common goals.

Coalitions with International Groups

Neither the toxic substances produced in the United States nor the health damage they cause stop at our country's borders. In fact, multinational corporations, aided by lax government regulation, spread their dangerous products and wastes around the world. The following examples illustrate the magnitude of the problem:

- U.S. companies such as Dow Chemical, Eli Lilly, DuPont, Monsanto, and Chevron sell pesticides banned in this country to Latin American buyers. Cancer-causing DDT, for example, is still available in many South American nations. To increase their yield, cotton planters in Guatemala spray U.S.-made pesticides thirty, forty, or even fifty times a season. In the mid-1970s, average blood levels of DDT among Guatemalans living in agricultural areas were among the highest in the world.[74]
- In 1977, the U.S. Occupational Safety and Health Agency banned the pesticide DBCP after it was shown to cause sterility among men working in pesticide manufacturing. When Shell Chemical and Dow halted their domestic production of DBCP, a plant in Mexicali, Mexico, increased its output and shipped the product to U.S. firms still selling the pesticide for its limited legal uses.[75]
- In the 1950s and 1960s major United States pharmaceutical companies (Bristol Meyers, Warner Lambert, G. D. Searle, Smith Kline) moved to Puerto Rico in search of lower taxes, cheap labor, and the promise of lax enforcement of environmental regulations. Their production and waste disposal processes have caused health problems among their workers, reduced the availability of fresh water, polluted fishing grounds, and threatened to contaminate drinking water. In the last five years, many of these same manufacturers have left Puerto Rico for other Caribbean islands in the hope of finding still lower taxes and wages. In their wake, they have left high unemployment and a polluted environment.[76]
- In April 1977 the Consumer Product Safety Commission banned the use of Tris, a flame retardant manufactured by the Velsicol Chemical Company for use on children's sleepwear, because it was suspected of causing cancer. Clothing manufacturers then shipped Tris-treated pajamas out of the United States to sell in Third World countries.[77]

Why should U.S. environmentalists be concerned with this growing export of toxic substances and products? On the simplest level, it is morally repugnant to solve our own environmental problems at the expense of the rest of the world. U.S. wealth depends in large part on the exploitation of the natural resources and labor of the people of Asia, Africa, and Latin America. To compound the historical crimes of slavery, colonialism, and the continuing plunder of Third World countries with the poisoning of these peoples with American corporate leftovers and waste violates any standard of justice and decency.

But opposition to the export of hazardous material does not rest solely on moral arguments, for the poisons we ship abroad can return to haunt us. Phosvel, a neurotoxic pesticide produced by Velsicol, was banned in the United States in 1976, then sold to Mexico. Phosvel reentered the country on 800 million pounds of tomatoes imported from Mexico.[78] Unless dangerous products are totally banned, companies will look for ways to make a profit on them, and with a highly integrated world economy, a product sold in the Third World can easily return to the United States to contaminate our air, water, soil, or food.

In the long run, corporations will usually not develop safe technologies unless they are forced to do so. As long as it is cheaper to move factories to developing countries than to institute adequate environmental control measures, industry will pack up and leave when it is convenient. The U.S. asbestos industry's move to Brazil, Mexico, and Taiwan following the implementation of stricter regulations here illustrates this.

The most important reason for an alliance between U.S. environmentalists and people in developing countries is to gain political power to affect corporations. Multinational companies depend on their foreign investments for a major portion of their profits. Should the government or people of developing countries insist on strict regulations to protect their environment and health as a condition for doing business in their nation, they would have a powerful lever for changing corporate practices. A common program for international environmental standards by activists and unions in this country and abroad would increase the pressure on corporations even more.

A few tentative steps toward an alliance have already been taken. For example, in 1979 a Colorado waste disposal firm offered the African nation of Sierra Leone $25 million a year to accept hazardous wastes from the United States. In another case, a New Jersey newspaper reported that a former state politician was negotiating a deal to dump hazardous wastes on the western shore of Haiti, one of the poorest nations in the world. In both these situations, adverse publicity, protests by environmental groups, and international pressure halted the projects.[79]

In 1981 the Institute for Food and Development Policy, an independent group based in California, released a study on corporate dumping of pesticides. Its report, *Circle of Poison*, helped spur the creation of the Pesticides Action Network, an association of nongovernmental agencies from around the world that within a year recruited more than eighty organizations and individuals in the United States alone.[80] Its members included church, environmental, labor, farmworker, development policy, and consumer groups. At a meeting of the network, Anwar Fazal, president of the International Organization of Consumer Unions, called pesticide dumping "a human rights issue because pesticides are killing people who are growing export crops—food you eat here in America." He vowed that the Pesticides Action Network would "build enough momentum around the world that the 'circle of poison' of hazardous pesticides manufactured in the developed world and dumped in the third world will eventually be broken."[81] In 1982 the Reagan administration proposed new export regulations that would relax the requirement that governments be notified when banned toxic products are shipped to them. His proposal raised such a storm of protest from national environmental groups and others that its implementation was at least temporarily stalled.[82]

These efforts by American activists to halt the export of hazardous materials are important precedents, but much more needs to be done to make the challenge to multinational corporations more than symbolic. The following suggestions describe some activities that might help strengthen an international environmental movement.

Consumer Boycotts, Shareholder Resolutions, and Public Education Campaigns

Some corporations have dumped chemicals in residential areas of the United States, shipped hazardous products overseas, and poisoned their workers at home and school. An international boycott of such companies could help educate the public about the global dimensions of the problem. The worldwide campaign against the Nestlé Food Corporation, because of its promotion of infant formula in developing countries, provides one model for such a project. That campaign was instrumental in persuading the World Health Organization to pass a code calling for strict regulation of the advertising of infant formula.

Support for National Liberation Movements

Independence movements in Third World countries have often taken a strong stand against corporate plunder. In Puerto Rico, for example, the nationalist movement has opposed exploitation of the island's nickel, cobalt, iron, and oil resources by the United States and other foreign countries for two reasons. First, they believe that the wealth generated by Puerto Rico's natural resources should benefit residents of the island rather than shareholders of mainland corporations. Second, they fear that companies will mine or process substances in ways that damage the health of islanders. Public control of the resources, nationalists believe, will increase the chances for environmentally sound development.

By supporting the Puerto Rican independence movement, mainland environmental activists contribute to the fight against pollution on that island. Similarly, solidarity work here in the United States in support of popular movements in Central America is one way the people in this country can strengthen forces willing to take a stand against multinational interests that overuse pesticides, contaminate water, and sell dangerous products.

Technical Assistance and Information to Activists and Governments in Developing Countries

Many of the hazardous substances that threaten people in developing countries are produced here in the United States. Activists and scientists from developed countries need to organize networks of communication that will allow us to share what we know about toxic chemicals with those in Third World nations. The Pesticide Action Network is one example of such an effort. Some researchers and activists have provided direct technical assistance. For example, a team of U.S. occupational health specialists went to Nicaragua to help investigate and treat an outbreak of mercury poisoning. International cooperation also helps establish ongoing links between environmental activists that can then be used to address subsequent problems.

Support for National and International Regulations to Control the Export of Hazardous Technology or Products

As the wealthiest nation in the world, the United States sets the standard for environmental protection. By insisting that corporations use the best available technology to control their pollution, U.S. activists establish a precedent that other nations can follow. Moreover, if domestic regulations were also to apply to the foreign operations of U.S. companies, environmentalists would remove one of industry's most powerful ploys used to continue polluting, the threat to move to a country with less regulation. Obviously, the application of U.S. regulations to overseas subsidiaries or the establishment of international regulations is currently beyond the reach of the environmental movement. In the long run, however, such safeguards offer the only hope for preserving a world whose air, water, food, and soil can sustain human life. Moreover, such regulations would directly benefit workers in the United States and abroad because it would reduce the possibility of job blackmail.

The potential for coalitions between environmentalists and others is not limited to those we have discussed here. The safe-energy movement has already addressed many of the health consequences of various energy sources. Organizations fighting for lower utility rates can also look at the environmental impact of utility companies' practices. Farmers' groups have joined the fight against corporate exploitation of mineral resources, and, less often, the campaigns against widespread spraying of pesticides. Consumer organizations have lobbied against hazardous products and for stricter environmental regulations. Some of the references listed at the end of the book describe the issues that might unite (and those that divide) environmental activists and these other groups.

This chapter has discussed how the environmental movement can expand its practices to include new issues and new constituencies. The ultimate goal of this expansion is to build a movement that can identify and correct the root causes of the environmental problems we face today. The last section describes some of the issues that activists need to consider as they seek to organize broader coalitions.

Building Coalitions: A Strategic Perspective

The success of the environmental movement in the next decade will depend on its ability to reach out to new constituencies. Yet, as we have seen, the road to effective coalitions is fraught with obstacles. Not only do real political differences separate environmentalists from each of the groups described in this chapter, but corporate interests work hard to accentuate and magnify these differences. Activists need a practical perspective on coalitions that will help both to address the day-to-day questions that come up in an organizing campaign and to build a movement that can become a significant political force.

Organizers initiate coalitions for a variety of reasons. The most common justification is pragmatism: "Since we are not strong enough to win by ourselves," ask these coalition builders, "whom can we convince to join us to increase our power?" The usual *quid pro quo* in this type of effort is "We'll go to your demonstration (or sign your petition or lobby with you) if you'll do the same for us." Other activists decide whom to ally with on moral grounds. "We think your cause is just, so we will join you," they assert. Groups joining a campaign for moral reasons do not necessarily expect anything in return for their support. Still another rationale for forming coalitions is "We're fighting the same enemy, so let's pool our efforts." In this more sophisticated argument, the success of one group in a coalition will benefit the other participants by weakening the power or credibility of the common enemy.

Each of these motivations for linking with other groups creates unique advantages and disadvantages. In the short run, the pragmatic coalition builders have the easiest task. The organizers of one group propose a deal to another that is either accepted or rejected. Neither party is considering a prolonged union, so political differences between the two are not of primary importance. As an organization collects a file of political debts and loans, it can pick and choose partners for a specific campaign as the need arises. Since the benefits of cooperating are usually clear to both sides, members are often willing to live up to the terms of the agreement.

In the longer run, however, coalitions built solely on a pragmatic foundation can topple. The failure to address the real differences between groups can lead to the sudden emergence of fierce political debates. The incentive to deal with more basic issues—for example, class, gender, or racial differences between the two groups—may not be there. Moreover, what seems like pragmatism to some activists may be opportunism to others. Feminists' anger at labor leaders' proposal to invite right-to-lifers to join the Coalition for the Reproductive Rights of Workers illustrates how "practical" decisions to expand a coalition can in fact fracture it.

Moral solidarity, the decision to take action because it seems the right thing to do, is a powerful political stimulus. It contributed to white participation in the civil rights movement and fanned the student movement against U.S. involvement in the war in Vietnam. Among environmentalists, moral arguments have led to action on the export of hazardous materials and have helped to move the antinuclear movement from fighting for safe energy to waging peace. When the thirst for social justice motivates activists to take on new issues, it is harder to quench their ardor with anything less than success. And to invite other organizations to join a battle because it is right gives the initiator of the coalition a powerful moral legitimacy that contrasts with the atmosphere of political deals created by more pragmatic organizers.

But moral solidarity has its drawbacks. For better or worse, most people will not engage in political action for long without some direct benefit to themselves. Except for the small group of people already committed, the rewards from ethical solidarity are often insufficient to sustain the coalition for the long run. Moreover, moral arguments can exclude precisely those groups most important to bring into environmental coalitions. Ethical solidarity has its strongest appeal to the white middle class people whose basic material needs are being met. For those who are still trying to meet their needs for decent housing, adequate food, employment, and so on, the request to make sacrifices (of time or energy) for someone else might seem élitist. This might explain the hostility that conservationists sometimes encounter from blue-collar workers or people of color.

Coalitions based on the desire to fight a common enemy combine some of the advantages of the other two types. On the one hand, both allies stand to benefit from the joint efforts. On the other hand, the alliance grows out of a shared analysis of the causes of a problem. Thus the common program has a political rationale that can be described and, if necessary, debated. These two characteristics give these coalitions both the immediate appeal of the pragmatic model and the principled unity of the ethical one.

Coalitions based on shared long-term goals also create a context for the kinds of debates and discussions that will be necessary if the groups are to overcome their differences. If activists understand why they need to work together, they can find the energy to address the issues that divide them. As groups gain experience and trust in working out mutual problems, they begin to develop a common politics that allows still closer cooperation.

The analysis of the causes of environmental health hazards in the United States presented in this book offers still another reason for pursuing coalitions based on a common opponent. The primary causes of pollution in this country are the social and political imperatives of modern capitalism and so the real solution to environmental problems is to transform a system that puts profit ahead of human needs. Clearly such a project is beyond the scope of the environmental movement. But if the various groups now pursuing their objectives independently (the labor movement, feminists, people of color, the peace movement) could define common goals, their collective likelihood of success would be greatly increased. Local alliances around specific issues are one of the best ways for these groups to gain the experience that would be necessary to launch national coalitions.

As we have seen in the case histories, coalitions can begin for a variety of reasons. But how they start is less important than the direction they go in. Even the most opportunist of alliances can be the starting point for an ongoing dialogue. Even the most principled of agreements can break apart in the crucible of practice.

Activists need to approach coalitions with a clear head and realistic goals. By defining mutual expectations explicitly, by articulating the basis for unity, by establishing mechanisms for resolving differences or even for dissolving the partnership, environmental groups increase their chances for building coalitions that can survive and grow.

Successful coalitions require organizers with skills and experience in bringing people together to work for common goals. They also require a political vision of the kind of movement we are trying to build. In the last chapter, I will describe one such vision and suggest how we can begin to implement it.

10. TOWARD A NATIONAL ENVIRONMENTAL MOVEMENT

Case History 1: The Japanese Environmental Movement

The environmental movement in Japan was founded by the victims of its postwar industrial expansion.[1] In 1956 a mysterious disease of the central nervous system was discovered in the Minamata Bay area.[2] An investigation by scientists determined that it was the mercury-contaminated wastes dumped into the bay by the Chisso Corporation, a major chemical manufacturer, that were causing the disease. By 1971 more than 120 cases had been reported, including 22 infants who were born with mercury-induced cerebral palsy or mental retardation. Some experts estimated that as many as 10,000 people had been exposed to dangerous levels of mercury and were thus at risk of developing the disease.

In the late 1950s a large industrial complex was built in Yokkaichi in the Mie Prefecture. By 1960 an epidemic of asthma had hit the city, and its cause was soon pinpointed as the sulfur emissions of factories in the complex. The condition became known as Yokkaichi asthma.[3]

Still another disease emerged in the Toyama Prefecture. In 1946 Dr. Hagino Noboru noticed a new syndrome characterized by severe nerve pain, softening of the bones, skeletal deformations, and extreme vulnerability to fractures. His research led him to conclude that the condition, labelled itai-itai disease, was caused by cadmium poisoning.[4] By 1972 Dr. Hagino alone had seen 280 cases of itai-itai disease, and at least 120 people had died from it.

In 1971, 512 of the victims won Japan's first major damages suit against a polluter. The suit had been instituted several years earlier by the Council for Countermeasures Against Itai-Itai Disease, a local organization of victims of cadmium poisoning, assisted by sympathetic labor unions and the Japanese Socialist and Communist parties. They charged the Mitsui Mining and Smelting Company with discharging toxic cadmium waste into the public water supply. Their contention was supported by Dr. Hagino's research and by a 1966 government study that confirmed his findings.

In 1970 residents of Toyama used a new referendum procedure to attempt to force the local legislature to pass stricter pollution regulations. More than 21,000 signatures were collected, requiring the governor to call a special session to consider the new proposals. Although the new rules failed to pass, the Council for Countermeasures learned valuable lessons in political mobilization. In municipal elections in Toyama and in another nearby town in 1971, mayoral candidates of the ruling conservative party were defeated by the reformist candidates of the Socialist party, in part because of the Socialists' support for the itai-itai victims.

Stunned by the size of the damages awarded in the 1971 court case, Mitsui decided to appeal. But the higher court only compounded the company's problem: it awarded even larger damages to the plaintiffs and ordered Mitsui to pay their opponents' court costs as well. To avoid further liability, Mitsui also agreed to compensate all certified victims who had not participated in the lawsuit and to institute cleanup procedures. Most important, it signed a "pollution prevention" agreement with set guidelines for its future operations.

In the following years, several other groups, including those representing victims of Minamata disease and Yokkaichi asthma, used the compensation and pollution prevention precedents won against Mitsui to win legal actions against corporate polluters.

In 1971 Prime Minister Tanaka proposed an ambitious plan for industrial development in rural areas. One such project was to be a petrochemical complex, including an oil refinery and a chemical plant, on the shores of Shibushi Bay in the Kagoshima Prefecture. The local government requested the national environmental agency to remove a nearby forest from its protected status, which outraged the local population and sparked a massive protest movement, known as the Liaison Movement to Prevent Petroleum Development in the Osumi Peninsula.[5] The protestors opposed the petrochemical complex on several grounds. First, they charged that it would pollute their air and water, leading to an increase in illness. Second, they feared the plant would destroy *local fishing, the area's major source of income, and the potential for tourism.* They argued that the industry's promise of jobs was illusory because highly trained technicians from the city, rather than local residents, would be hired. Finally, they believed that the heavy industry and the influx of urban workers would threaten the region's distinctive culture, values, and scenery.

The Liaison Movement was able to organize strong opposition to the plan. In late 1971 residents sent a petition with thirty-four thousand signatures to the national government, demanding that the complex not be built. In some small towns nearly 80 percent of eligible voters signed. The Communist and Socialist parties, as well as labor unions and fishermen's cooperatives, also supported the movement's demands, and municipal workers in one city staged a sympathy strike. In 1972 the local government twice had to call in riot police to remove protestors from the legislature, where they had gathered to demand action to halt the plant.

The growing movement eventually forced the government first to scale down and then to abandon the Osumi project. In 1978 and 1980 the local government tried to institute a smaller complex, but the Liaison Movement continued its efforts to block construction.

In recent years, Japanese environmental activists have focused on opposing that country's ambitious nuclear power plans—by 1979 there were fifteen nuclear power plants in operation.[6] The Japanese peace movement had long opposed nuclear weapons, a position seared into public consciousness by the atomic bombs the United States had dropped on Hiroshima and Nagasaki. Its support added strength to the foes of nuclear energy. Protests against the nuclear-powered ship Mutsu illustrate the depth of this opposition.[7] In 1972 the Mutsu was ready to set sail, but demonstrations and court orders kept it in port for two more years. In August 1974 it finally prepared to go to sea, only to find that local fishermen had mobilized a fleet of three hundred fishing boats, which surrounded the vessel. On shore, twenty thousand demonstrators cheered their success. When a tugboat tried hauling the Mutsu through the barricade, the fishermen chopped through its dragline with hatchets.

As a result of the protest the national government promised to find a new port for the Mutsu and not to make any repairs on its reactor in its home port. Yet seven years later the government had

still failed to find a city that would accept the ship. Public opinion polls conducted before and after the Mutsu incident showed a
dramatic increase in public opposition to nuclear power.

Rapid industrialization and urbanization in Japan after World
War II created a host of environmental hazards. The government's
firm commitment to U.S.–sponsored economic growth made it
unresponsive to citizen complaints about pollution and health
hazards. The lack of established mechanisms by which people
could participate in making economic decisions contributed
further to their alienation from the political system. The result
was that between 1965 and 1975 a grass-roots movement against
local hazards became a national political force.

In a study of the Japanese environmental movement, political
scientist Margaret McKean describes some of its accomplishments.[8] First, the movement changed individuals. Tens of
thousands of people demonstrated, petitioned, or campaigned for
the first time. According to McKean's survey of activists, most of
them had not previously been politically active. Their actions
changed their political attitudes, beliefs, and behavior. Some
switched allegiance from the conservative Liberal Democratic
party to the more progressive Socialist or Communist parties.
Many became increasingly critical of industry's promises to build
a new, more prosperous Japan. Perhaps most important, participants gained a new sense of personal independence and power.
In a society that strongly emphasized obedience to established
authority, this was a crucial first step toward building a movement capable of a sustained fight against the power structure.

On the community level the environmental activists set a
precedent for mass participation in political decision making.
Through lawsuits, demonstrations, and legislative lobbying, environmental groups showed that it was possible to challenge government or corporate decisions and win. Their success in winning
the support of labor unions, left political parties, and occasionally even conservative officials demonstrated that it was possible
to build local coalitions that could exercise influence.

Although the Japanese environmental movement does not have
national organizations similar to the Sierra Club or the National
Wildlife Federation in the United States, local action contributed
to the passage of legislation that is significantly stricter than any
in the United States. As early as 1969 the government was forced

to establish a compensation program for victims of pollution, and by 1979 over seventy-three thousand people had been certified as suffering from pollution-related diseases.[9] Further, local governments often passed laws stricter than the national ones. Throughout the 1970s the number of environmental ordinances passed, pollution-control personnel hired, and violators arrested increased steadily.

Finally, the Japanese environmental movement challenged the ideology of government and industry that claimed that continued economic growth offers the sole route to prosperity and material well-being. A movement led by victims of this "prosperity" dramatically demonstrated its consequences. The combined actions of the environmental, student, and peace movements and certain sectors of organized labor forced people to begin debating the political and economic directions they wanted Japan to pursue. While the international depression of the early 1980s slowed the momentum of Japan's insurgent movements (by raising the fear of unemployment), their legacy continues to provide a foundation for future struggles.

Case History 2: The Greens, West Germany

In the 1983 West German national elections the Greens, a coalition of environmentalists, peace activists, feminists, and leftists that had organized as a party only in 1979, won 2 million votes (5.6 percent of the total), sending twenty-seven of its candidates to the Bundestag, the German equivalent of the U.S. House of Representatives.[10] Although the conservative Christian Democratic party won control of the federal apparatus, the Greens' victory assured that the antimilitarist, ecological, and democratic aspirations of Germany's insurgent social movements would receive a hearing at the national level. Greens had also won seats in six of West Germany's eleven state legislatures, including West Berlin.

How did this victory come about, and what can the Greens hope to accomplish in the years to come? As in Japan, the West German environmental movement is rooted in the failures of postwar political and economic development. In the 1950s and 1960s a rapidly expanding economy led to prosperity, but also to massive pollution. To fuel continued economic expansion, an ambitious nuclear power program was initiated, and by 1978 twelve plants were generating power and several others were under construction.

But the worldwide economic crisis of the late 1970s hit West Germany hard. By 1983, 2.5 million people were out of work, a record 10 percent of the workforce. The model welfare state administered for more than a decade by the Social Democratic party could no longer meet the needs of an economically strapped population, and its only response to the growing discontent was to build new jails, hire more police, and pass repressive laws to "combat terrorism." It also adopted an increasingly anti-Soviet and pro-U.S. foreign policy, replacing its earlier cautious pursuit of détente with Eastern Europe.

It was in this climate that the Greens became a national political force. Many of its members had become politicized by actions of the militant antinuclear movement. For example, opponents of a nuclear power plant in Grohnde, in the north, built an "anti-atom village" near the site.[11] The occupiers constructed a playground, friendship houses, and an information center. The police laid siege for two months, and then 1,300 officers used tanks and buses to break up the village. In another action three years later, 5,000 activists staged a four-week live-in at Gorleben in the east at the proposed site of an underground waste disposal site for radioactive by-products from reactors throughout Western Europe.[12] After four weeks the largest squad of police deployed in Germany since Hitler (8,000 strong) attacked the protestors, as well as many reporters and photographers. Peaceful demonstrations protesting police brutality erupted in twenty-five cities across the nation. From these and other incidents a national network of environmental activists emerged.

But opposition to the militarization of West Germany and specifically to the stationing of U.S.–made cruise and Pershing II missiles became the issue that drew increasing numbers to the Greens. Public opinion polls show that an overwhelming majority of Germans oppose the deployment of U.S. missiles there.[13] In June 1982 more than half a million people gathered in Bonn to protest Reagan's visit there; throughout 1983 demonstrations and civil disobedience to protest the missiles continued in West Germany.

The Green Party has worked hard to support local activism while at the same time maintaining a national political presence. One method for doing so was to use open general assemblies of the party as the final decision-making body. Hundreds, sometimes even thousands, of people meet to plan strategy. Leadership is rotated so that power does not become institutionalized, and candidates elected to local or national offices resign their seats halfway through their terms and are replaced by alternates. The salaries of representatives to state legislatures are divided among the officeholder, the party, and an "ecology fund" that is used to finance local projects.[14] The emphasis on process and grass-roots democracy attracted those who were disaffected with bureaucracy. "The Green Party grew," said Petra Kelly, one of its leaders, "because people streamed to it as to a therapeutic institution. This is the strength of the Green Party—no structure to suppress you."[15]

Within parliamentary bodies the Greens look for opportunities to raise radical issues. As one observer put it, "Electoral politics is used as a place to make waves."[16] For example, a Green Bundestag representative said that although he would generally abide by state secrecy provisions, he felt compelled to inform the public of "matters of vital interest to them—such as the planned sites of Pershing II nuclear missiles."[17]

Not surprisingly, the Greens' organizational and political choices have also created problems. Some activists opposed the move into electoral politics, believing it would undercut local activism. As West Berlin organizer Michael Lucas observed in 1978,

> In the upswing of electoralism, a large portion of the energy, the imagination and political work that would otherwise go into extending the grassroots organizational structure of the movement through direct action tactics will be channeled into campaigns to place radicals in parliament on a local and national level. As the entire movement is pulled further to the right by the more powerful rightist forces it will be much easier for the government to criminalize and politically isolate more radical tendencies within the movement.[18]

So far, however, the Greens' problem has not been co-optation. Their participation in the large and growing peace movement in Western Europe helps to balance their parliamentary activities, and the Christian Democratic victory makes a Green-Social Democratic alliance highly unlikely, since the Greens' radical stance on disarmament has made the Social Democrats wary of cooperation.

Until recently the Greens have not attracted much labor support. Their preoccupation with nuclear weapons and the environment has prevented them from addressing traditional labor concerns, while their ecological consciousness has occasionally alienated the unions. A Green member in the Baden-Wurtemberg state legislature, for example, claimed to be the parliamentary representative of all of Germany's toads, who required protection from use in radioactive experiments.[19]

At its January 1983 congress, however, the Greens began to work with the main West German trade union federation to develop a program for social reconstruction. In addition, a few Green activists have long been involved in labor politics. Rainer Trampert, for example, a party spokesman, is a member of the elected workers' council at the Hamburg Texaco plant where he works.[20]

Another problem facing the Greens is the heterogeneity of its constituencies. Although its slogan is "unity within diversity," diversity sometimes triumphs. Whether the many shades of Green that make up the party can hold together long enough to influence West German military, economic, and environmental policies remains an open question. Sectarian squabbles have crippled popular German movements before.

With the deployment of the U.S. missiles in early 1984, the West German peace movement lost some of its momentum, at least temporarily. But the Greens have already made important contributions. More than any political group in an advanced capitalist country, the Greens have developed a critique that links opposition to military buildups, corporate destruction of community life, environmental degradation, and authoritarian government. Their broad popular support testifies to the growing disenchantment with the fruits of capitalist development, while their analysis of the relationship between the oppression of people and the destruction of nature has provided a powerful tool for building a mass movement. As the Greens and their successors develop a more systematic program for rebuilding society, they may play a growing role in West Germany.

In Japan, community struggles headed by the victims of environmental pollution led to a national movement that forced the government to institute new regulations and made corporations compensate those they poisoned and adopt new pollution control technology. By forming coalitions with labor groups, the peace movement, and progressive political parties, Japanese environmentalists opened a national debate on the costs and consequences of capitalist expansion.

In West Germany, environmental activists created an organization that could both win representation in the national legislature and help to sustain militant local fights against environmental hazards and military projects. Moreover, their organization, the Green Party, became a meeting place for diverse social movements—feminists, peace activists, left political parties, conservationists, and anarchists. The Greens' determined opposition to military expansion linked them to a broader European movement. At the same time, the Greens used both local and national forums to put forth a critique of West German society and to articulate a vision of a more democratic, decentralized, and ecologically sound society.

Important differences in political history, electoral systems, and social consciousness distinguish the United States from West Germany and Japan, and attempts to import blueprints from another country seldom succeed. Moreover, the ultimate impact of German and Japanese activists has yet to be determined. Nevertheless, the accomplishments of environmental movements in these advanced capitalist countries have important lessons for the U.S. movement. In both Japan and West Germany, activists succeeded in three critical areas. First, they created channels to link local campaigns with national issues. Second, they influenced the national political discourse on the relationship between capitalist expansion and its environmental consequences. And third, they succeeded in reaching out to new political constituencies and thus building a broader force for change.

These three accomplishments constitute the critical challenges for the U.S. environmental movement in the 1980s. Can we in this country achieve these goals? What have we accomplished so far, and what obstacles block further progress? Answers to these questions require an assessment of the strengths and weaknesses of the two fundamental components of the environmental movement in this country, the grass-roots community groups, and the national conservation organizations.

Local Environmental Organizations

In the last five years, grass-roots environmental groups have won many concrete victories. They have prevented the construction of nuclear power plants and toxic waste facilities. They have won passage of state and local legislation such as right-to-know bills, moratoria on nuclear power plant construction, and bans on aerial herbicide spraying. They have forced corporations to install pollution control equipment, allow their plants to be inspected, and withdraw dangerous products from the market. They have made the government clean up abandoned waste sites, enforce regulations more vigorously, and monitor health problems more closely. While we will never know how many lives have been saved, how many cases of cancer prevented, or how many genes protected from damage, these community groups have undoubtedly helped to safeguard the health of the public.

Local groups have also educated the public about the nature and content of environmental pollution. According to a 1980 public opinion poll, 90 percent of the respondents said they would not want a nuclear power plant near their home. More than 40 percent said they would not want to live within five miles of a toxic waste dump.[21] Ten years ago almost no one would have worried about these hazards. Through flyers, demonstrations, and media attention, community groups have taught millions of Americans about threats to their health, corporate greed, inadequate government protection, and the need for action. The local groups' roots in the neighborhood and their firsthand encounters with toxics have made them compelling educators.

One result of this education has been that hundreds of thousands of people have been mobilized to action in their own communities. Some, like Lois Gibbs, had been housewives uninvolved in local politics. Others, like Helene Brathwaite in Harlem or Edwina and Gene Cosgriff on Staten Island, had been active in school or church groups. Still others, like the Native American women in WARN or the leaders of the right-to-know campaign in Philadelphia, were already sophisticated activists. Whatever their route of entry, however, those who formed or joined community groups developed new political skills, mastered immense amounts of technical information, and created new forms of social action. The leaders and members of these groups constitute a cadre of experienced organizers and activists who can be the backbone of any national environmental movement.

Like their counterparts in West Germany and Japan, local groups have been able to draw in people not previously involved in environmental issues. Ranchers in Utah and Wyoming, farmers in Minnesota, poor rural blacks in North Carolina, urban blacks and Hispanics in New York City, Native Americans in South Dakota and New Mexico, union members in California and Pennsylvania, working-class women in New Jersey, middle-class suburbanites on Long Island and in New England—all these and more have fought to protect their communities from corporate pollution and inadequate government regulation. Few social issues in this country have aroused so many people across class, race, and gender lines. Few have created such a potential for broadly based progressive coalitions.

Local environmental groups have also posed a threat—however indirect—to corporate control of social planning. U.S. corporations have always been able to pass most of the costs of environmental pollution on to the public—to workers, consumers, and taxpayers. They have also been able to plan production, transportation, and waste disposal with minimal consideration of the environmental impact of such decisions. Inadequate government regulation, lax enforcement of existing pollution control laws, and the lack of public planning all help make such corporate control possible.

Community environmental groups have challenged the consequences of this. Through demonstrations, lawsuits, ordinances, referenda, and media coverage, they have forced corporate managers to consider the effects of their decisions on the surrounding neighborhoods. Every delay a group wins, every concession it wrests, costs the corporation money and diminishes corporate control. At some point, corporations will decide that it is easier and cheaper to give in to some demands than to suffer the delays, the product liability suits, or the bad publicity that local opposition can create. Dow Chemical's 1983 decision to withdraw from the market the herbicide 2,4,5-T illustrates this point.[22] By forcing industry to calculate the economic and political costs of irresponsible environmental practices, activists have provided some inducement for greater attention to health and safety.

It would be a mistake, however, to overestimate the relative strength of community groups—theirs is still the power of the flea that bites the elephant. Despite their accomplishments, they can never constitute a sufficiently strong political force to prevent environmental pollution. For one thing, the corporations operate nationally, even internationally. If one plant is forced to respond to community demands to clean up, the corporation can move the dirty operation to another state or another country. If one local government bans the transportation of radioactive material, the nuclear power industry can use its influence in Washington to overturn the ban. If one pro-environmental candidate is elected to Congress, corporate donors can ensure that five others more sympathetic to their view are elected elsewhere. In short, environmental pollution is a national problem; its prevention requires national solutions.

A movement limited to local groups tends to engage in defensive battles. Local groups learn of a hazard after it has done its damage; they are faced with a *fait accompli* that has to be reversed. An EPA study of community opposition to hazardous waste sites explained how industry and government resist such after-the-fact efforts:

> Experience has shown that it is much more difficult for public opposition to shut down an operating facility than to prevent a facility siting. A major reason for this is that, unless the facility is violating state regulations in its operations, the state regulatory agency will probably feel obliged to defend its regulatory process and thus the site.[23]

Defensive battles lead to piecemeal solutions. The limited power of the community groups forces them to accept compromises. Moreover, corporations with national offices and highly paid lawyers and public relations specialists learn from their experiences in fighting local groups. They develop new strategies to co-opt opponents, subvert the intent of regulations, or change unfavorable laws. Concerned residents, on the other hand, have a much harder time learning from other struggles and thus often make the same mistakes.

The case histories in this book show how local activists have been able to build organizations that can win victories and continue to monitor government or corporate performance. But they also show that such victories require an immense amount of time and energy. For each group that succeeds, many others fail or win some immediate gains, then dissipate. A movement that depends on each local organization's sustaining itself will lose many activists who become frustrated and impatient; as a result, only the most determined organizations will survive. The environmental movement can ill afford this loss of its human resources.

The decentralized, amorphous character of the groundswell of opposition to environmental pollution has its strengths; its flexible tactics keep its opponents guessing; the lack of national leaders makes co-optation or repression difficult; and its local roots nourish democratic participation and facilitate communication with a range of constituencies. But there is a price to be paid for this diversity. The movement lacks a coherent program: some people become active to keep a hazard out of their neighborhood; others hope to encourage industry to act more responsibly; still others seek a society in which human needs take precedence over profits. Class and race differences among activists contribute to these varied political perspectives. The very heterogeneity of the movement—in one sense a strength—makes it difficult to agree on priorities or strategies.

Corporations do not suffer from such ideological disagreements. Large and small oil companies may differ on how much lead they should have to remove from gasoline, and pollution control device manufacturers may disagree with petrochemical companies on the enforcement of environmental regulations, but basically all the corporations agree on a political program that minimizes government involvement in environmental protection and guarantees their ability to make production decisions. This consensus makes it easier to play on the differences among environmentalists. Some of their opponents can be bought off or appeased, others ignored, still others threatened or sued. As long as each environmental group follows its own course, corporate polluters can divide them and conquer them.

Defensive struggles, fuzzy politics, and isolated groups lead to another problem. A movement that seeks to involve masses of people in changing institutions, policies, and values needs a vision that links the often frustrating day-to-day work with an image of a better world for now and the future. Such a vision helps to draw people into action and sustains them over the long run. The early civil rights movement in the South, for instance, had the vision of a society in which blacks had dignity and equal rights. In West Germany and Japan, environmentalists have begun the task of defining an alternative to the bureaucratic, centralized corporate state.

In the United States, the grass-roots movement has an implicit vision of a society in which people have the right to a healthy environment but it has not been able to articulate what such a society should look like, and it has not clearly defined how it will muster the political power necessary to bring about change. This makes it difficult for people to move beyond fighting to protect their own backyards.

In sum, local environmental groups have provided some of the crucial building blocks for a national movement: public awareness of the problems, a varied repertoire of effective strategies and tactics, a network of experienced, battle-tested organizers and activists, and the first steps toward broader coalitions. But their lack of a coherent program or a national strategy has blocked them from having a still greater influence.

The National Environmental Organizations

The second major force for environmental protection in the United States is the multitude of national organizations—the Sierra Club, the National Wildlife Federation (NWF), the Wilderness Society, Environmental Action (EA), Friends of the Earth, the Audubon Society, and others. Although each of these groups has a different priority, constituency, and strategy, they are all committed to the same broad goals, and in the last few years they have begun to collaborate more actively. They have also, thanks to President Reagan, taken on a more openly political mission than ever before. This has enabled them to play the national role that has so far eluded grass-roots activists.

To provide a better understanding of the strengths and weaknesses of the large environmental organizations, I will begin with sketches of four of these groups.

The Sierra Club was founded in 1892 by John Muir, a conservationist who had lived alone in the Sierra Mountains for many years and was dedicated to protecting the unspoiled wilderness. Until the late 1960s, the club's primary mission was to safeguard the wilderness against commercial industrial development. Its members, at first mainly Californians, were middle- and upper-class outdoors people. Club activities included lobbying, legal action, and hikes and wilderness trips.

By 1970 the Sierra Club had 113,000 members.[24] The burgeoning ecology movement raised new environmental issues—nuclear power, pollution, toxic chemicals, overpopulation—that the club was quick to become involved in. Although its priority was still the national parks and its emphasis was still on Washington-based activities, some of its members were active in the community environmental struggles described in previous chapters.

By the late 1970s the club had begun to devote more of its resources to electoral politics. Carl Pope, a Sierra staffer, explained the rationale for this new emphasis:

> When politicians can be fairly certain that their actions will meet with no great public outcry, the corporations by default gain a virtual monopoly on power, a monopoly which is an overwhelming barrier to achieving environmental goals. The only way to end that monopoly is to find some new way to hold politicians accountable.[25]

President Reagan's environmental policies have helped to push the Sierra Club still further in this direction. In October 1981 the club presented Congress with a petition bearing 1.1 million signatures calling for the ouster of Reagan's secretary of the interior, James Watt. Nearly 100,000 new members joined in 1982 alone. Of the 158 candidates endorsed by the club in the elections that year, 80 percent won.[26] Sierra Club Executive Director Michael McCloskey vowed to expand the club's emphasis in this sphere: "We will spend the eighties learning the ropes of electoral politics," he said. "We can't match the corporations' PACs [political action committees] in dollars, but we intend to make it up in shoe leather."[27]

With 4.3 million members, the National Wildlife Federation is both the largest and one of the most traditional of the conservation groups. Founded in 1835 as the American Wildlife Institute, the NWF was financed by the gun industry and other corporate donors who wanted to "manage" wildlife so as to ensure continued game for hunters.[28] But the destruction of wildlife habitat by corporate development moved the federation to take a more activist stance. In 1980, for example, it published a handbook for citizens on how to protect communities against toxic substances. Its advice was to work within existing channels, but people were urged to take action to prevent the repetition of "environmental horror stories."[29]

President Reagan's policies moved the NWF further in politics, leading it to join some of the coalitions that were lobbying in Washington. However, the NWF defined a more cautious path for itself than did such groups as the Sierra Club. Jay Hair, a zoologist who is the NWF's executive vice-president, said he hoped to work with the administration "to see that the environment is not the scapegoat for the nation's economic problems."[30] He also called for a policy of "corporate détente" to help end traditional conflicts between industry and environmentalists.

The Sierra Club and the National Wildlife Federation represent the range of political perspectives among the older conservation organizations. The next two, Environmental Action and Friends of the Earth, are the progeny of the more recent ecology movement.

Environmental Action is perhaps the most explicitly anticorporate of all the national groups. Among its founders were organizers of the first Earth Day, in 1970. Unlike the Sierra Club and the NWF, EA does not have active local chapters but has a small Washington-based staff that travels throughout the country, lobbying, training, and providing technical assistance to many different local groups. Environmental Action staffers have also been active in state fights against nuclear power and acid rain and for bottle bills and other conservation and recycling efforts.

EA was a prime mover in such national coalitions as the Clean Air Coalition, which in 1982 successfully fought industry and Reagan administration efforts to weaken federal air pollution laws. It was also a partner in an "indictment" of the Reagan administration in which ten organizations charged the president with "engaging in a wholesale give-away to private interests of our most precious natural resources."[31] EA has also played an active role in the National Campaign to Stop the MX, the missile system that it claims "poses the greatest threat both to general disarmament efforts and to the environment of the American West."[32]

From its inception, EA has been active in electoral politics.

Beginning in 1970 it targeted for defeat in each national election the "Dirty Dozen," a group of twelve congressmen with the worst environmental voting records for that session. The group also identified what it called the Filthy Five—Amoco, Dow Chemical, Occidental Petroleum, Republic Steel, and Weyerhauser—which EA called "big polluting companies that contribute heavily to political candidates," and also targeted recipients of this money for defeat.[33]

After the Republican victory in 1980 produced a Congress virtually impervious to lobbying efforts, EA set up its own political action committee (PAC) "to help win back some seats for Mother Earth."[34] In 1981 EA/PAC mobilized volunteers and dollars in New Jersey's state elections to help eight candidates it had endorsed—six won. In 1982 seventy-five of the ninety-eight congressional candidates endorsed by EA won.

Friends of the Earth was established in 1969 by David Brower after he had resigned as executive director of the Sierra Club because of organizational and political differences. The most important of the substantive conflicts concerned nuclear power. At its annual meeting in 1969 the Sierra Club had endorsed the construction of a nuclear power plant at Diablo Canyon, south of San Francisco. The club argued that this site was better than a previously proposed one in a more ecologically fragile area. Brower and his supporters opposed all nuclear power plants, a position subsequently adopted by Friends of the Earth. (Later, the Sierra Club also came out against nuclear power.[35])

Other major issues Friends of the Earth addresses include wilderness preservation, wildlife protection, and toxic chemicals. It has affiliates in twenty-six nations, making it an important advocate for international environmental action. It actively opposes the nuclear arms race and in a 1983 full-page advertisement in the *New York Times*, the group explained why it had taken up this issue:

> Until recently we were content to work for our usual constituency: life in its miraculous diversity of forms. . . . But the nuclear war contemplated by the U.S. and Russia would kill life of *all* kinds, indiscriminately, on a scale and for a length of time into the future that is so great that it qualifies as the major ecological issue of our time. In fact, from an ecological point of view, the war has already begun as the mere preparations for it are causing illness and depriva-

tions to humans, bringing harm to the planet's life-sustaining abilities, and casting a deadly pall over life on Earth.[36]

These four sketches illustrate how diverse the national organizations are. What can they contribute to a movement for a healthier environment, and what factors limit their usefulness?

Without a doubt these groups have put environmental protection on the national political agenda. The legislative victories of the early 1970s—the Clean Air and Water acts, the National Environmental Policy Act, the Toxic Substances Control Act—could not have been passed without their active lobbying and campaigning. And while the Reagan administration succeeded in weakening federal involvement in environmental regulation, the lobbying and advocacy of groups like the Sierra Club, the Audubon Society, and the NWF Foundation have helped to minimize the damage. As Russell Peterson, president of the Audubon Society, has put it, "a lot of the things the Reagan administration set out to do have been stopped dead in their tracks."[37] Further, these groups' constant attacks on the policies of James Watt and EPA Administrator Anne Gorsuch Burford helped to create the political climate that forced Reagan to accept Watt's resignation and to dismiss EPA's entire top management.

The national environmental organizations have also provided valuable support to grass-roots struggles. The right to a public hearing, the mandate for environmental impact statements, and the right to institute legal challenges that these organizations helped win in the first generation of environmental legislation have in turn helped local activists in their own battles.

In recent years the national organizations have also given direct technical assistance to community groups. For example, the Environmental Defense Fund, a group dedicated to legal efforts to clean up the environment, sent a community organizer to Love Canal. The Natural Resources Defense Council, another group that works mainly in the courts, and the Sierra Club have filed numerous legal briefs in support of local opponents of polluters. Regional chapters of national organizations have often provided volunteers who have become important members of community groups. For newly formed neighborhood environmental associations, these resources often make the difference between survival and rapid extinction.

The national organizations' renewed interest in electoral politics has added an additional pressure point for environmental concerns. In 1982 these groups spent $2 million on their election campaigns.[38] After being reelected to a congressional seat in 1982, California Democratic Representative Phillip Burton noted that if the Sierra Club and the Friends of the Earth "had done the same thing for my opponent [that they did for me], he would have won."[39] This ability to make the difference between victory and defeat gives environmental groups an added tool to keep politicians accountable. National electoral activity also provides activists with a way to address concerns larger than their particular local hazard. Campaigns can become a forum for getting disparate community groups to agree on common goals. The federal elections in West Germany, for example, forced the Greens to come up with a program that could unite their diverse supporters.

In sum, then, these organizations have become effective advocates for environmental protection. Their sophisticated lobbying, their success in building new coalitions, and their electoral victories meant that by 1982 they were being considered—by as conservative a voice as *Business Week*—"a national power capable of influencing economic, foreign and social policies."[40] Warning that the environmental movement has "established a substantial beachhead in American electoral politics," the magazine raised the spectre of a movement that might emulate the Greens' "persistent tendency toward negativism—if not nihilism."[41]

At present *Business Week*'s fears of a U.S. movement critical of capitalism are probably exaggerated. Unlike the Green Party, U.S. national conservation organizations have neither direct representation in Congress nor a critique of society accompanied by an alternative program. Further, their influence in Washington has been established at the price of grass-roots militancy. For one thing, they have chosen to emphasize electoral, legislative, and legal strategies rather than local organizing, so that their battles usually take place in institutions where industry and government can easily outspend and outmaneuver them. Industry plays a dominant role in defining the rules of the game, and environmental groups, with less funds and political power, are forced to compromise—to accept partial limitations on exposure to a toxic substance, for example. These compromises in turn legitimate the political process, which is stacked against ordinary folk. "See," say the politicians and industry officials, "the system does work; citizens can participate, and now only 10,000 people per year will die of pollution-related lung cancer instead of 20,000."

The point is *not* that environmentalists should never compromise. But the national organizations' decision to play by the rules written by corporations and the government ensures that they will win only limited victories and that the political debate will center on the pros and cons of modest reforms of existing social and economic institutions. If, as I have argued in earlier chapters, these institutions are *themselves* the cause of our environmental problems, then a movement that accepts their inevitability will never succeed. Here the contrast with the Greens in West Germany and the victims' groups in Japan is crucial: they raised more fundamental questions about democracy and economic growth, challenging the idea that capitalist expansion benefits most people—the supposed exchange for environmental destruction.

The national organizations' emphasis on electoral politics illustrates the limits of working within the system. The Reagan administration's environmental policies led most of these groups to endorse Democratic candidates, although some supported liberal Republicans such as Senator Robert Stafford in Vermont (a champion of clean air). But while some of the Democratic victors in 1982 congressional races voted the right way on important environmental issues, they were unable to reverse the basic thrust of the Reagan administration. Similarly, the conservation groups' public prestige—as well as their electoral victories—helped to convince President Reagan to drop Anne Gorsuch Burford and James Watt. The president was worried about the impact of their poor public images on the 1984 elections. But their replacements, while more skilled at public relations, made few substantive changes in pro-industry policies at EPA and the Department of the Interior. Moreover, simply changing the administration in power in Washington will not solve the environmental problems in the United States. Although President Carter appointed environmentalists to important EPA posts, his Democratic administration first proposed the MX missile, offered to relax EPA emission control standards on certain automobiles, and badly bungled the Love Canal disaster. As long as Republicans and Democrats agree that increased industrial growth is necessary to maintain a healthy economy, that corporations have the right to shape that economy—and as a result, the environment—as they choose, and that the public health benefits of environmental protection must be balanced against their costs to corporations, the parties' tactical differences on how much money to give to EPA or what regulatory loopholes to close are relatively minor.

Yet most environmental organizations have been reluctant to move beyond the Democratic or Republican parties. In an interview in early 1983, Sierra Club director Mike McCloskey said, "I doubt if there will be any need to think in terms of a separate political party concerned with the environment. In a two-party system, a third party has almost no chance of success in congressional elections."[42]

McCloskey is probably accurate in his assessment of the difficulty of short-term victories for a third party. But the decision to work within the two-party system relegates environmentalists to a narrow choice of political options. Since both parties depend on corporate contributions and reflect corporate ideology, it is unlikely that either will challenge corporate control of economic planning and regulatory policy. National electoral activity may lead to marginal differences in enforcement and spending for environmental protection, but it is unlikely to lead to substantive changes that will prevent environmental degradation and threats to our health. Thus, it seems clear that if national organizations put all their energy and resources into electoral politics, they will be leading the movement down a dead-end street. On the other hand, as the Greens have shown, using the electoral arena as a forum for public education, as a place to make waves, can give environmentalists a way to influence the national political discourse. So far, however, the conservation organizations have not used their political influence in this way.

The national organizations have sometimes thwarted rather than helped struggles at the local level. David Foreman, a militant ecologist who became a staff person for the Wilderness Society, describes how he was called on "to keep the grass roots in line" so as to prevent activists from filing a lawsuit against an adverse Forest Service ruling on forest protection.[43] The national organization did not want to alienate the senators and representatives with whom they hoped to negotiate a legislative compromise on the issue.

Foreman charged that the "top conservation staffs of these

groups fretted about keeping local conservationists (and some of their field staffs) in line, keeping them from becoming extreme or unreasonable, keeping them from blowing moderate national strategy on a variety of issues."[44]

In part, the national organizations' desire to work within the system stems from honest differences of opinion. Many of their leaders believe that national lobbying and electoral efforts are the best hope for protecting the environment. They see tactics or strategies that jeopardize national victories as quixotic and ineffective. In part, however, their perspective reflects their own socioeconomic position. Directors and members of the Sierra Club, for example, are mostly white middle-class professionals or small business owners. The Natural Resources Defense Council's thirty-five-member board of trustees in 1981–1982 had ten lawyers, seven academics, and three foundation directors—only five of whom were women. It is thus not surprising that these organizations are unwilling to advocate political programs that would seriously challenge the status quo.

Another problem the national environmental organizations pose to those who favor a stronger, broader movement is that they continue to emphasize conservation rather than public health. Although groups like the Sierra Club, the Audubon Society, and Friends of the Earth have made significant progress in expanding their mandate to include such issues as toxic chemical hazards and air pollution, their hearts are still often in the wilderness. For example, in 1982 the Sierra Club's magazine had twice as many articles discussing wildlife as human health issues.

This focus on conservation alienates potential supporters in the labor unions and among blacks and other minorities, who feel that a movement that puts animals and trees ahead of jobs and health does not represent their interests. As a result, politically experienced and powerful groups are lost to the environmental movement. At the same time, the emphasis on conservation attracts wealthy individuals who favor protection of national parks or endangered species as long as this does not threaten existing economic and social institutions. With their prestige and financial connections, these individuals often become leaders of the national organizations—putting the leadership in the hands of people whose own socioeconomic interests are diametrically opposed to those of poor and working people.

Again the contrast with Japan is instructive. There, victims of pollution were the movement's leaders, putting public health at the center of the debate. As a result, labor unions, community groups, and the left political parties joined the environmental crusade.

In the long run, a U.S. environmental movement would gain from understanding the interrelationships between conservation and human health. What damages people also harms plants and animals, and vice versa. On a larger scale, as World Wildlife Fund vice-president Thomas Lovejoy observed, "flora and fauna represent 10 million successful solutions to a wide array of biological problems, each of which is relevant to our existence in some way."[45] The peregrine falcon alerted us to the danger of DDT contamination, and poisoned fish have provided early warning of water pollution by mercury, polychlorinated biphenyls (PCBs), and dioxin. As we have seen earlier, both the ecofeminists and Native American groups have linked the destruction of nature with damage to human health. Their efforts, as well as those of Japanese and West German activists, provide models of how the conservation organizations could broaden their perspective. Their success in doing so will depend on rank-and-file members insisting that organizations respond to the needs of poor and working people.

In the decades to come the environmental movement will be battling for two fundamental rights: the right to live in an environment that does not damage health and the right to participate in making decisions about the environment in which one lives. Winning these rights will challenge other currently accepted behaviors such as the ability to use one's property without restriction (the "right to pollute") and the ability to plan with little or no public accountability.

This coming battle will pit environmental activists, concerned residents, conservationists, and workers against those who own and run U.S. industry. Can environmentalists hope to win this lopsided struggle? Can a movement be built that has the ability to challenge one of the most powerful political forces on earth?

A strategy for a successful movement must meet several criteria. First, it must be able to link the efforts of grass-roots activists with those of the national environmental organizations. Local groups provide the militance, the ties to masses of people, the commitment to public health, and the democratic spirit that a strong movement requires. The conservation organizations, in turn, have technical expertise, political experience, and financial resources. While there are important differences in philosophy and goals between the two, their common interest in environmental protection and their complementary strengths and weaknesses make cooperation in their mutual self-interest.

Second, a successful movement must create opportunities for forging coalitions with the groups described in the previous chapter. By themselves, environmentalists are unlikely to be able to make major political changes. If they join their efforts to those of labor unions, people of color, feminists, and peace groups, their chances for success are far greater. An environmental movement will have to learn how to define its goals so as to bring new people into the struggle.

Third, a successful movement needs a vision of a society that values human needs, health, and the environment more highly than corporate profit and control. Defensive battles against one polluter or another cannot by themselves create a movement; only by providing its supporters with a view of a society with a different spirit and values and institutions can a movement sustain their commitment.

In my opinion, only a socialist movement can meet these criteria. Socialism offers a critique of capitalism, an explanation of its inherent tendencies toward socially destructive practices, a rationale for coalitions and alliances, and a vision of a society based on human needs. Although socialists have not always been sensitive to environmental concerns, socialist theory provides an intellectual and political framework for building a movement that can challenge the *causes* of environmental degradation.*

But further action to protect the environment cannot await the development of a strong socialist movement. What is needed is a strategy that can move us from where we are to where we need to be—a strategy that will put the issue of public ownership and control of social and natural resources on the agenda while at the

---

*While socialism is a necessary condition for an ecologically sound society, it is not a sufficient one; witness the environmental problems in the Soviet Union. Environmental health hazards in the USSR are a product of the social, political, and economic decisions made there in the post-World War II period. Soviet environmentalists have already begun the task of addressing these conditions.[46]

same time providing the opportunity for concrete victories that can build a movement.

One element of a political strategy is a program, a statement of specific goals. A program gives activists the opportunity to say, "This is what we stand for; this is what we are trying to accomplish."

A program for an environmental movement will help activists in several ways. It will:

- Embody a vision of a society that values human health and the environment above corporate profit and control.
- Encourage dialogue and debate on the causes of environmental pollution and on the direction a movement seeking to prevent it should take.
- Offer specific demands that could bring women, workers, minorities, peace activists, and others into environmental struggles.
- Identify targets for action at local, regional, and national levels and suggest links between local and national issues.
- Lead to concrete victories that demonstrate the power of collective action.
- Help to build a unified movement that can make lasting changes.

Such a program can emerge only out of the experiences of the movement and out of its activists' critical analysis of that experience. I suggest here a preliminary program for an environmental movement. My goals are to articulate some of the issues that have already brought environmental groups together and to stimulate the debate that must precede a collectively defined national program.

A National Program for
A Healthy Environment

The environmental movement has raised three fundamental questions:

1. What political rights do individuals need in order to win and maintain a safe and healthy environment?

2. What are government's responsibilities for environmental protection?
3. What are industry's responsibilities for safeguarding the environment and public health?

The program that follows spells out how environmentalists have answered these questions. It seeks to consolidate the varied demands that have emerged in local struggles into a coherent strategy for a healthier, more just society.

Democratic Rights

1. Community residents and workers have the right to know the names, properties, and known health effects of all toxic substances produced, stored, transported, or disposed of in their environment.
2. Community residents and workers, their organizations and their designated experts, have the right to inspect any facility that is suspected of damaging health or the environment.
3. Community residents and their organizations have the right to participate in any deliberations that may lead to decisions that could threaten their neighborhood.
4. People have the right to refuse any project they believe will damage their health or that of future generations.

Access to information is the prerequisite for meaningful participation in the planning process. Obviously, implementation of these four rights raises thorny moral, legal, and practical problems. What kind of evidence as to risk is acceptable? What political structures can be used to exercise these rights? (A few possibilities: ballot referenda, citizen advisory boards with real power, elected environmental boards.) At present community organizations and labor groups are fighting for these rights on a piecemeal basis—at the municipal, state, and federal levels, as well as with individual corporations. Their experiences will help activists plan effective strategies that can be generalized to wider situations.

Environmentalists undoubtedly will disagree, both among themselves and with industry and government, as to the best methods to guarantee these rights. Some individuals or communities will undoubtedly be able to violate these rights by acting on spurious assessments of risk or will be able to protect their narrow self-interest at the expense of a greater social good. But every human right engenders conflict. For more than two hundred years, U.S. citizens have been debating what freedom of speech means and how to balance this right against other social goals. Despite these disputes and problems, almost everyone agrees that freedom of speech is a right worth preserving.

A movement that poses the right to a healthy environment as its goal can win widespread support from the people of the United States. Moreover, the debate on this right will raise fundamental questions about who owns the air, water, and soil; who decides what risks are acceptable; and who plans how to meet the needs of present and future generations. The right to a healthy environment is the cornerstone of a program for a national movement.

Government Responsibilities

The public perception that the government has failed to fulfill its responsibility to protect the environment has precipitated many community and national struggles. A successful program must specify exactly what government agencies must do to meet their obligations to the public.

1. Existing government laws must be strictly enforced by means of regular inspections and monitoring, and stiff penalties for violators. Regulatory agencies at the local, state, and national levels must have the resources they need to fulfill their legal obligations. Transferring money from the bloated military budget to environmental protection would improve public health in two ways: it would lead to a cleaner environment and it would reduce the risk of war.
2. The government must establish national priorities for environmental health research together with funding and other incentives. A research agenda must focus on how to prevent or minimize human exposure to previously produced hazardous substances and how to develop rapid and effective procedures for screening new substances for toxicity. Industry must pay for government-regulated testing of the products it expects to produce and profit from.
3. While better research and strict enforcement of existing regu-

lations will significantly improve environmental quality, they are not sufficient. The long-term goal of public policy must be to eliminate human exposure to toxic substances. Until it can be proven beyond a reasonable doubt that a toxic substance has a threshold level below which no biological changes take place, the goal must be to prevent *any* exposure to that substance.

Zero human exposure is, of course, a radical demand, but if this is made the goal of public policy many of the current problems with setting and enforcing regulations would be minimized. If a toxic substance is detected in an environment where humans can be exposed, there is a violation. The Delaney Amendment to the U.S. Food and Drug Act embodies this concept by banning the presence of any level of carcinogens in food for human consumption. Achieving the goal of zero human exposure would stop the endless and frustrating scientific task of attempting to set individual exposure standards for thousands of dangerous substances. It would also eliminate many of the expensive, time-consuming procedures that now plague both regulators and generators of toxics. It would prevent the kind of "experimentation" that has led industry to expose tens of millions of people to untested products.

Obviously, zero exposure cannot be implemented all at once. Scientists, industry, and the public must hammer out a schedule for achieving this goal. Further, some exceptions may be necessary—there are perhaps a few toxic substances whose benefits to society so outweigh their risks that the public will want to continue producing them. In the long run, however, by changing manufacturing processes, substituting safe products for hazardous ones, testing materials before marketing them, and so on, such exceptions will become extremely rare.

By establishing zero human exposure as our goal, we will fundamentally alter the equation for environmental decision making. The bottom line will become the health of the environment and of the people it supports both now and in the future. *Cost-benefit* analysis, which values corporate profits more highly than public health, will be replaced by *cost-effectiveness* analysis. Industry will be able to decide which pollution control methods are most effective for eliminating human exposure, not whether it will control its emissions.

Corporate Responsibility

While environmental activists have more often attacked government than industry for causing environmental problems, the root source of most pollution is industrial and agricultural production.

Corporations need to be confronted with their responsibility for environmental protection; they can no longer simply plan for their own benefit, at public expense.

1. Corporations must compensate individuals and communities for the damage they do to property or health. Frank Kaler, a New Jersey man whose family's water supply was contaminated with toxic waste, explained the rationale for such compensation: "If a man came into my backyard with a gun and tried to hurt my children, I could shoot him," he said. "But if someone poisons my backyard and hurts my kid, I can't do a damn thing."[47] Compensation will serve two purposes: it will act as a deterrent to pollution and provide some moral and financial relief for those who have had to bear the expense of illness, disability, or death. Where the particular corporation responsible for the pollution cannot be identified, the government must tax classes of industry (e.g., the petrochemical or nuclear power industries) to establish compensation funds for those hazards.

2. Industry must dispose of all toxic materials at the site of production. This will eliminate the need to transport these materials, encourage recycling and redesign of the manufacturing process to minimize waste, and force companies to pay their own disposal costs. It will also provide an incentive for creating products with minimal hazardous by-products. In those few cases where on-site disposal is impossible, other incentives and penalties for minimizing waste must be developed.

3. Industry (with government oversight) must prove a substance safe beyond a reasonable doubt before it can be produced and marketed. The burden of proof must thus be shifted from the victim of poisoning to the poisoners. It is sound public policy to assume a substance toxic until proven innocent. Industry must bear the cost of this testing; government must specify the criteria that govern evidence of safety.

4. In the transition from today's unsafe environmental practices

to safer ones, employers must not be able to dismiss or demote employees and use those savings to pay for meeting environmental regulations. Companies that have created unsafe environmental conditions have profited from these actions; they cannot pocket the profits and shift the burden of meeting regulatory requirements onto workers (who are themselves probably also poisoned) by closing plants, reducing staff, or moving abroad.

In those few cases where an entire industry might need to be shut down (for example, the nuclear power industry), workers must receive paid job retraining and retain all pay and seniority rights at public expense, funded by a tax on the industry responsible for damaging the public health (in this case, the electrical utilities). It is not fair to shift these expenses onto consumers or taxpayers, who are not involved in making the original decisions.

Such a policy will have two important effects. First, it will set a precedent that industries that operate in wanton disregard of public health and safety cannot shift the subsequent cost of their policies onto their workers, consumers, or the taxpaying public. Second, it will allow workers to participate in environmental and occupational health struggles without the fear of losing their jobs.

5. U.S. environmental regulations must be made to apply to the operations of U.S. corporations anywhere in the world. As the wealthiest and most industrially developed nation, the United States can afford to set a standard for other nations to emulate. And as a nation that has grown rich by exploiting human and natural resources around the globe, it has a moral obligation not to poison the lands and people who have contributed—often under duress—to this wealth. A shift of regulatory responsibility from Manila, Tegucigalpa, or Lagos to Washington will make it harder for multinational corporations to browbeat the governments in developing countries to let them get away with environmental murder.

As we have seen, the application of U.S. regulations to all corporate operations will also benefit people in this country. Substances like DBCP and DDT will not return to haunt us after having been used in the Third World on foods that are then sold in this country. Workers and communities will be less vulnerable to environmental blackmail. If a corporation knows that the same standards apply in the United States, Mexico, Brazil, and Taiwan, the incentive to move south will be reduced.

Conclusion

There is nothing new in this program: every point has been raised already in some political struggle. But what it offers is a framework for building a movement that can get to the roots of this country's environmental problems. It combines minor reforms, some of which have already been partially achieved (the right to know), with major challenges to corporate power. Some of its points can be easily won in the next few years; others will need a long, hard fight. As a whole, the program provides a tool for beginning to build a stronger, more cohesive, more militant movement. How can it be used?

Community groups can use relevant sections to focus their local struggles. An organization that is fighting to close a toxic waste facility, for example, can insist on having the rights to know, inspect, and refuse. Their goals should include not only closing down a dangerous facility but also the right to monitor the company's promises and to acquire information they can use in subsequent struggles. Similarly, residents negotiating for the cleanup of an abandoned toxic dump site will insist that the long-term goal should be zero human exposure to the dump's toxic chemicals, not simply a temporary cleanup that will cause new problems in the future.

When groups around the country press for similar demands, they increase the pressure for tougher regulations at the national level. When dozens of community organizations seek and win (as they have in Massachusetts and Texas) the right to inspect a factory or monitor a corporation's concessions, they create a climate in which such rights can be guaranteed by national legislation.

The program also provides a concrete focus for local, regional, and national coalitions. Coalitions of environmental organizations, peace groups, labor unions, and public health professional associations have already begun to fight for stricter regulations and better enforcement. Right-to-know bills have led to community-labor alliances in many cities and states. In several regions, coalitions or networks have brought groups together to shape common goals and strategies, as well as to share information. In the future, new coalitions may coalesce around the issue of corporate responsibility, targeting notorious polluters (such as Velsicol Chemical Company or Occidental Petroleum) and demanding that their victims across the country receive compensation for their losses.

In the electoral arena, a strong national movement unified around a program can negotiate with politicians from a position of relative strength. Its principles and positions are explicit, so candidates can be supported or opposed based on their willingness to adopt all or part of the program. Moreover, an ideologically united movement can hold those it elects accountable—politicians who fail to keep their promises will immediately lose the support of a significant bloc of voters. A program will not eliminate the problems of a two-party system described earlier, but it can allow a more principled participation. Moreover, the program makes clear that the goal is not simply to change the people in the White House or Congress but rather to force substantive changes in policy.

A program also provides a starting point for independent political groupings (such as the Citizens party, the Association of Community Organizations for Reform Now [ACORN], and Fair Share) to join with environmentalists to run their own candidates for office. Environmental groups can make the acceptance of all or part of the program a condition for joining such third-party efforts.

In the long run, however, it seems clear that U.S. environmentalists will need a single national organization to represent them. Only by concentrating our numbers and unifying our political perspective can we hope to change an economic system that puts a low priority on health and the environment. It is premature to predict whether such an organization will be a network of existing environmental groups, a federation of progressive organizations that addresses many issues, or a formal independent political party. The Green Party, with its emphasis on local democracy, provides us with one model and the even more loosely connected victims groups in Japan with another. In this country, the movement now lacks the ideological consensus and political infrastructure to launch a national organization. But the debate on a political program will encourage activists to consider future directions for the environmental movement.

Despite these problems, in the last few years environmentalists around the country have taken important steps toward building a more unified movement. Some examples:

- Community environmental groups in North Carolina, Tennessee, Wisconsin, New Jersey, and other states have formed regional or statewide coalitions to press for stricter regulation, right-to-know laws, mandatory citizen involvement in the siting of toxic waste facilities, and statewide superfunds that tax petrochemical industries in order to pay for cleanup of spills or dump sites.
- National environmental organizations have achieved an unprecedented unity in combating Reagan's environmental policies and they have helped to make environmental health an important issue on the nation's political agenda.
- Many groups not previously active in environmental issues—the National Women's Health Network, the National Association for the Advancement of Colored People (NAACP), the United Church of Christ's Commission for Racial Justice, ACORN, Massachusetts Fair Share, disarmament groups, safe-energy groups, and scores of others—have taken important stands on local and national environmental problems, indicating the potential for coalitions in the years to come.
- National organizations such as Environmental Action, the Environmental Defense Fund, the Urban Environment Conference, and the National Citizen's Clearinghouse on Hazardous Waste, formed by the veterans of Love Canal, have started training programs for local activists. Such training plus the dozens of regional and national conferences held in the last three years have helped to build the leadership and the political and technical sophistication the movement needs.

All these activities have brought the environmental movement to a crossroads. Environmentalists now have the attention of the American people. Where will we take this new interest? Down one road is business as usual. Local activists fight local battles, each group defending its backyard against one threat after another. National organizations work for reform, never stepping over the invisible line that might lead those in power to sever their ties with them. Environmentalists stick to fighting for the environment, peace activists to fighting the military, and labor to fighting for its own economic needs.

Down another, more rocky road lies a militant movement that dares to challenge the fundamental assumptions of U.S. capitalism. It is a movement that demands the right to a healthy environment and the right to decide one's own destiny. It includes blacks, whites, Native Americans, men and women, working people, city dwellers and farmers, students, and senior citizens. It is a movement that is willing to debate political principles and to test new strategies and tactics. It is a movement based in neighborhood groups, community coalitions, regional networks, and national organizations. It is a movement with a coherent analysis of the causes of environmental problems and with a political program for their solution. It is a movment that seeks the power to make this country safe for life—now and for the generations that will follow.

Those who make up the environmental movement now, both the veteran organizers and those who have recently become active, must choose which of these paths to pursue. On this decision rests the future of our movement, our health, and our environment.

REFERENCES

1: The Corporate Assault on Health

1. The case history is based on the following sources: "Together We Can Do It: Fighting Toxic Hazards in Tennessee: Interview with Nell Grantham," *Southern Exposure* 9, no. 3 (Fall 1981): 42–47, and author interview with Nell Grantham, September 1981.
2. C. S. Clark et al., "An Environmental Health Survey of Drinking Water Contamination by Leachate from a Pesticide Waste Dump in Hardeman County, Tennessee," *Archives of Environmental Health* 37 (1982): 9–18.
3. "Together We Can Do It," p. 45.
4. Ibid., and telephone interview with Nell Grantham, December 1983.
5. The case history is based on the following sources: Kris Melroe, "On the Edge of Extinction," *Off Our Backs* 9, no. 5 (May 1979): 8–9; Women of All Red Nations (WARN), "Pine Ridge Reservation Health Study," February 1980, Mimeographed; Lorelei Means and Madonna Gilbert, "Radiation: 'Dangerous to Pine Ridge Women,' WARN Study Says," *Akwesasne Notes*, Spring 1980, pp. 22–23; WARN, "Report to the Russell Tribunal: Continued Genocide of the Lakota People: Corporate Contamination of their Water," October 1980, Mimeographed; and telephone interview with Jacqueline Huber, attorney for WARN, April 1982.
6. The case history is based on the following sources: *New York Times*, 17 December 1980, p. B1; 10 March 1981, p. 82; 1 August 1980, pp. 27–28; *Daily Journal* (Newark), 23 March 1982; Coalition for a United Elizabeth (CUE), *On Cue*, June 1979, pp. 1–2; May and June 1980, pp. 1–3; July 1980, pp. 1, 2, 4; January–February 1982, pp. 3–6; and interview with CUE Associate Director Jacinta Fernandes and toxics staffperson Gloria Davis, March 1982.
7. *New York Times*, 1 August 1981.
8. Fernandes interview.
9. Debra Lee Davis and B. H. Mayer, "Cancer and Industrial Chemical Production," *Science* 206 (1979): 1356–57.
10. Milton Weinstein, "Cost-Effective Priorities for Cancer Prevention," *Science* 221 (1983): 17–23.
11. Council on Environmental Quality, *Environmental Quality: 10th Annual Report* (Washington, DC: Council on Environmental Quality, 1979).
12. Frances Moore Lappé and Joseph Collins, *Food First* (Boston: Houghton Mifflin, 1977), p. 175.
13. Charles Komanoff and Eric van Loon, " 'Too Cheap to Meter' or 'Too Costly to Build'?" *Nucleus* (Union of Concerned Scientists) 4 (1982): 3–7.
14. *New York Times*, 16 May 1982, sect. 3, p. 1.
15. *New York Times*, 2 August 1982, p. D8.
16. W. Keith Morgan et al., "Respiratory Disability in Coal Miners," *Journal of the American Medical Association* 243 (1980): 2401–4.

17. Appalachia—Science in the Public Interest, Citizen's Blasting Handbook, 1978, p. 30.
18. Highlander Research and Education Center (HREC), We're Tired of Being Guinea Pigs! (New Market, Tenn.: Highlander Center, 1980), p. 3.
19. Ibid.
20. Kai Erikson, Everything in Its Path (New York: Simon & Schuster, 1976).
21. Philip Landrigan et al., "Epidemic Lead Absorption Near an Ore Smelter: The Role of Particulate Lead," New England Journal of Medicine 292 (1975): 123–29.
22. Washington Post, 7 October 1979.
23. New York Times, 25 April 1983, p. A10.
24. Leslie J. Freeman, Nuclear Witnesses (New York: W. W. Norton, 1981), p. 144.
25. Los Angeles Times, 17 September 1979.
26. WARN, "Pine Ridge Reservation Health Study."
27. Cited during "Four Corners: A National Sacrifice Area?" WNET-TV, 14 November 1983.
28. J. C. Marx, "Low-Level Radiation: Just How Bad Is It?" Science 204 (1979): 160–64.
29. General Accounting Office, Problems in Assessing the Cancer Risks of Low-Level Ionizing Radiation Exposure, Comptroller General's Report to Congress, EMD 81-1, 1981.
30. Anna Gyorgy, No Nukes (Boston: South End Press, 1979), p. 46.
31. New York Times, 11 April 1982, p. 19.
32. Rachel Scott, Muscle and Blood (New York: E. P. Dutton, 1974), pp. 10–11.
33. New York Times, 9 February 1982, p. A14.
34. Advisory Committee to the Office of the Secretary, Department of Health, Education, and Welfare ("Rall Committee"), "Health and Environmental Effects of Increased Coal Utilization," Federal Register 43 (16 January 1978): 2229–40.
35. S. C. Morris et al., Data Book for Quantification of Health Effects from Coal Energy Systems, Brookhaven National Laboratory Regional Energy Studies Program, Contract EV 76-C-07-0016 (Upton, N.Y.: U.S. Department of Energy, 7 May 1979).
36. New York Times, 29 June 1983, pp. A1, A12.
37. U.S. Environmental Protection Agency, Acid Rain, vol. 1 (Washington, D.C.: EPA, April 1980).
38. H. M. Braunstein et al., Environmental Health and Control Aspects of Coal Conversion, vols. 1 and 2 (Oak Ridge National Laboratories, prepared for Energy and Development Administration, April 1977).
39. New York Times, 28 July 1981, p. B4.
40. New York Times, 13 September 1981, p. 48.
41. Nuclear Information and Resource Center (NIRC), Nuclear Waste: Where It Comes From (Washington, D.C.: Nuclear Information and Resource Center, 1981).
42. Gyorgy, No Nukes, p. 55.
43. NIRC, Nuclear Waste: Where It Comes From.
44. Leonard Solon, "Public Health Perspective in the Highway Routing of Radioactive Materials through Populated Areas," Statement to U.S. Department of Transportation, Material Transportation Bureau, 29 November 1977.
45. HREC, We're Tired of Being Guinea Pigs!, p. 10.
46. Andrew Marino and Robert Becker, "Hazard at a Distance: Effects of Expo-

sure to the Electric and Magnetic Fields of High Voltage Transmission Lines," *Medical Research Engineering* 12 (1977): 6–9.

47. Nancy Wertheimer and Ed Leeper, "Electrical Wiring Configurations and Childhood Cancer," *American Journal of Epidemiology* 109 (1979): 273–84.

48. National Institute for Occupational Safety and Health (NIOSH), *President's Report on Occupational Safety and Health* (Washington, D.C.: Department of Health, Education, and Welfare, 1972).

49. NIOSH, *National Occupational Hazard Survey*, vol. 3 (Washington, D.C.: U.S. Government Printing Office, 1977).

50. Council on Environmental Quality, *Environmental Quality—11th Annual Report* (Washington, D.C.: Council on Environmental Quality, 1980).

51. Lester B. Lave and Eugene Seskin, *Air Pollution and Human Health* (Baltimore: Johns Hopkins University Press, 1977).

52. *New York Times*, 1 August 1980, pp. 27–28.

53. Carl M. Shy, "Testimony before Subcommittee on Health and Environment," U.S. House of Representatives, Clean Air Act Amendments of 1977, Part 1, 8–11 March 1977, pp. 244–62.

54. *New York Times*, 28 March 1982, p. 26.

55. *New York Times*, 5 April 1982, p. B6.

56. *New York Times*, 16 May 1982, p. 47.

57. *New York Times*, 20 May 1982, p. A22.

58. Reported in *New York Times*, 30 November 1983, p. A1.

59. U.S. Congress, Senate, Committee on Commerce, Science, and Transportation, *Hazardous Materials Transportation's Regulatory Program* (Washington, D.C.: U.S. Government Printing Office, 1979).

60. National Transportation Safety Board, *Safety Effectiveness Evaluation—Federal and State Enforcement Efforts in Hazardous Material Transportation by Truck*, NTSB-SEE-81-2 (Washington, D.C.: U.S. Government Printing Office, 1981), p. 6.

61. Ibid.

62. U.S. General Accounting Office, *Program for Ensuring the Safe Transportation of Hazardous Materials Need Improvement* (Washington, D.C.: U.S. Congress, 1980).

63. U.S. Congress, *Hazardous Materials Transportation*.

64. *In These Times*, 16–29 July 1980, pp. 7–8.

65. *New York Times*, 20 November 1982, p. A14.

66. *New York Times*, 13 October 1982, p. A13.

67. H. A. Anderson et al., "Household Contact Asbestos Neoplastic Risk," *Annals of the New York Academy of Medicine* 276 (1976): 311–23.

68. *New York Times*, 10 February 1983, p. D4; 1 July 1983, p. A6.

69. *New York Times*, 22 February 1982, p. A1.

70. *New York Times*, 23 May 1982, p. B2.

71. Quoted in *New York Times*, 1 September 1983, p. A15.

72. *New York Times*, 22 April 1983, p. B2.

73. Nicholas A. Ashford et al., "Law and Science Policy in Federal Regulation of Formaldehyde," *Science* 222 (1983): 894–900.

74. Harriet Rosenberg, "The Home Is the Workplace: Hazards, Stress, and Pollutants in the Household," in *Double Exposure: Women's Health Hazards on the Job and at Home*, ed. Wendy Chavkin, M.D. (New York: Monthly Review Press, 1984), pp. 219–45.

75. *New York Times*, 31 August 1983, pp. A1, A18.

76. "Health Effects of Construction through a Landfill," *City Health Information* [New York City Department of Health] 1/19 (1982): 1–3; Clark et al., "En-

vironmental Health Survey of Drinking Water Contamination"; Thomas Maugh, "Just How Hazardous Are Dumps?" *Science* 215 (1982): 490–93; *New York Times*, 8 July 1982, p. B4; 2 January 1982, pp. 1, 7; 28 December 1981, p. A16; William Lowrance, ed., *Assessment of Health Effects at Chemical Disposal Sites*, Proceedings of a Symposium held in New York City on June 1–2, 1981, by the Life Sciences and Public Policy Program of the Rockefeller University, pp. 3–22.

77. *New York Times*, 29 June 1981, p. A12.
78. Michael Brown, *Laying Waste* (New York: Pantheon Books, 1980), pp. 28–59.
79. Reported in *New York Times*, 17 March 1983, p. B14.
80. Reported in *New York Times*, 29 December 1982, p. A8.
81. *New York Times*, 2 January 1982, p. 7.
82. Philip J. Landrigan and Richard Gross, "Chemical Wastes—Illegal Hazards and Legal Remedies," *American Journal of Public Health* 71 (1981): 987.
83. Quoted in *New York Times*, 13 March 1983, p. F2.
84. Reported in *New York Times*, 28 December 1981, p. A16.
85. Ibid.
86. *New York Times*, 4 November 1981, pp. B1, B2.
87. Ibid.
88. *New York Times*, 8 April 1983, p. B3.
89. *New York Times*, 25 May 1982, pp. C1, C7.
90. Ibid.
91. *New York Times*, 19 May 1983, p. D2.
92. Rachel Carson, *Silent Spring* (Boston: Houghton Mifflin, 1962), p. 14.
93. H. P. Chase et al., "Pesticides and U.S. Farm Labor Families," *Rocky Mountain Medical Journal* 70 (1973): 27–31; Ephraim Kahn, "Pesticide-Related Illness in California Farm Workers," *Journal of Occupational Medicine* 18 (1976): 693–96; U.S. Department of Health, Education, and Welfare, *Report of the Secretary's Commission on Pesticides and their Relationship to Environmental Health* (Washington, D.C.: U.S. Government Printing Office, 1969).
94. HREC; *We're Tired of Being Guinea Pigs!*, p. 25.
95. Samuel Epstein, *The Politics of Cancer* (Garden City, N.Y.: Anchor Books, 1979), pp. 245–47.
96. T. H. Milvey and D. Wharton, "Epidemiological Assessment of Occupationally Related, Chemically Induced Sperm Count Suppression," *Journal of Occupational Medicine* 22 (1980): 77–82.
97. U.S. Environmental Protection Agency, *Report of Assessment of a Field Investigation of Six-Year Spontaneous Abortion Rates in Three Oregon Areas in Relation to Forest 2,4,5-T Spray Practices*, Epidemiologic Studies Program, 28 February 1979.
98. *Los Angeles Times*, 30 July 1979.
99. HREC, *We're Tired of Being Guinea Pigs!*, p. 26.
100. *Suffolk Times* (Greenport, N.Y.), 3 June 1982, p. 7.
101. *New York Times*, 1 August 1982, pp. LI-1, 4.
102. Reported in *New York Times*, 4 December 1983, p. 65.
103. *Oakland Tribune*, 4 July 1982.
104. *San Francisco Chronicle*, 9 July 1982.
105. R. Jeffrey Smith, "Hawaiian Milk Contamination Creates Alarm," *Science* 217 (1982): 137–40.
106. *Los Angeles Times*, 1 September 1980.
107. Reported in *New York Times*, 15 March 1983, pp. A1, D24.
108. Ibid.
109. Ellen Grzech, "PBB," in *Who's Poisoning America*, Ralph Nader, Ronald

Brownstein, and John Richard, editors (San Francisco: Sierra Club, 1981), pp. 60–84.

110. Mary Wolff et al., "Human Tissue Burdens of Halogenated Aromatic Chemicals in Michigan," *Journal of the American Medical Association* 247 (1982): 2112–16.
111. Barry Commoner, *The Closing Circle* (New York: Alfred Knopf, 1971), pp. 78–90; James Turner, *The Chemical Feast* (New York: Grossman, 1970).
112. National Action/Research on the Military Industrial Complex (NARMIC), *Pentagon Country* and *Makers of the New Generation of Nuclear Weapons* (Philadelphia: American Friends Service Committee, 1981).
113. *New York Times*, 8 February 1982, p. A14.
114. Reported in *New York Times*, 7 April 1983, p. A18.
115. *New York Times*, 18 March 1983, p. A30.
116. *New York Times*, 9 September 1981, p. A12.
117. Ibid.
118. *New York Times*, 13 September 1981, p. 69.
119. Quoted in *New York Times*, 8 February 1982, p. B2.
120. *New York Times*, 26 December 1982, p. 22.
121. Commoner, *Closing Circle*, pp. 151–52.
122. Barry Commoner, "The Promise and Perils of Petrochemicals," in *The Big Business Reader*, ed. Mark Green and Robert Massie (New York: Pilgrim Press, 1980), pp. 348–49.
123. Richard Kazis and Richard Grossman, *Fear at Work: Job Blackmail, Labor, and the Environment* (New York: Pilgrim Press, 1982), p. 204.
124. Cited in Larry George, "Love Canal and the Politics of Corporate Terrorism," *Socialist Review* 66 (1982): 27.
125. *Philadelphia Inquirer*, 23 September 1979; *New York Times*, 7 September 1981.
126. Cited in Commoner, *Closing Circle*, p. 259.

2: Science and Politics

1. The case history is based on the following sources: Michael Brown, *Laying Waste: The Poisoning of America by Toxic Chemicals* (New York: Pantheon, 1980); Lois Marie Gibbs, *Love Canal* (Albany: State University of New York Press, 1982); idem, "Lessons from Love Canal," speech at the Conference on Hazardous Materials in the Metropolitan Region, Columbia University, New York, N.Y., 13 June 1981; Adeline Gordon Levine, *Love Canal: Science, Politics, and People* (Lexington, Mass.: Lexington Books, 1982); R. Jeffrey Smith, "Love Canal Study Attracts Criticism," *Science* 217 (1982): 714–15; idem, "The Risks of Living Near Love Canal," *Science* 217 (1982): 808–11; Dr. Beverly Paigen, "Health Hazards at Love Canal," testimony presented to the House Subcommittee on Oversight and Investigation, 21 March 1979.
2. Levine, *Love Canal*, pp. 11–12.
3. Ibid., p. 28.
4. Ibid., p. 29.
5. Ibid., p. 37.
6. Paigen, "Health Hazards at Love Canal."
7. Levine, *Love Canal*, p. 93.
8. Ibid., p. 101.
9. Ibid., p. 132.
10. Ibid., p. 116.
11. Ibid., p. 133.

12. Cited in ibid., p. 139.
13. Ibid., p. 149.
14. Quoted in *New York Times*, 27 May 1980, p. A1.
15. Cited in Levine, *Love Canal*, p. 158.
16. Ibid., pp. 159, 165.
17. Ibid., p. 168.
18. Smith, "Love Canal Study Attracts Criticism" and "The Risks of Living Near Love Canal."
19. Smith, "Love Canal Study Attracts Criticism."
20. Ibid., p. 714.
21. Ibid.
22. Quoted in *New York Times*, 22 June 1983, p. B3.
23. *New York Times*, 2 October 1982, sect. 4, p. 7.
24. Thomas Maugh, "Chemical Carcinogens: The Scientific Basis for Regulation," *Science* 201 (1978): 1200–1205.
25. Bruce Ames et al., "Methods for Detecting Carcinogens and Mutagens with the Salmonella/Mammalian Microsome Mutagenicity Test," *Mutation Research* 31 (1975): 347–64.
26. Lorenzo Tomatis et al., "The Predictive Value of Mouse Liver Tumour Induction in Carcinogenic Testing: A Literature Survey," *International Journal of Cancer* 12 (1973): 1–20; Samuel Epstein, *The Politics of Cancer* (Garden City, N.Y.: Anchor Books, 1979), pp. 52–72.
27. F. W. Sunderman, "A Review of the Carcinogenicities of Nickel, Chromium and Arsenic Compounds in Men and Animals," *Preventive Medicine* 5 (1976): 279–94.
28. Maugh, "Chemical Carcinogens."
29. William Hines and Judith Randal, quoted in Epstein, *Politics of Cancer*, p. 52.
30. Ibid., p. 2.
31. *New York Times*, 2 November 1982, p. A25.
32. Epstein, *Politics of Cancer*, pp. 69–70.
33. Office of Technology Assessment, *Cancer: Risk Assessing and Reducing the Dangers in Our Society* (Boulder, Colo.: Westview Press, 1982), p. 67.
34. Richard Doll and Richard Peto, "The Causes of Cancer: Quantitative Assessment of Avoidable Risks of Cancer in the U.S. Today," *Journal of the National Cancer Institute* 66 (1981): 1191–1308.
35. Frederica Perera and Catherine Petito, "Formaldehyde: A Question of Cancer Policy?" *Science* 216 (1982): 1285–91.
36. Quoted in Epstein, *Politics of Cancer*, p. 57.
37. Departmental Task Force on Prevention, U.S. Department of Health, Education, and Welfare, *Disease Prevention and Health Promotion: Federal Programs and Prospects* (Washington, D.C.: U.S. Government Printing Office, 1978), p. 89.
38. *New York Times*, 26 June 1983, p. 22.
39. Toxic Substances Strategy Committee, *Toxic Chemicals and Public Protection* (Washington, D.C.: U.S. Government Printing Office, 1980), p. 64.
40. R. Jeffrey Smith, "A Battle over Pesticide Data," *Science* 217 (1982): 517–18.
41. Keith Schneider, "Faking It: The Case against Industrial Bio-Test Laboratories," *Amicus Journal* 4, no. 4 (1983): 15.
42. Ibid., p. 14.
43. Reported in *New York Times*, 12 May 1983, pp. A1, A21.
44. *New York Times*, 9 May 1983, p. A17.
45. Schneider, "Faking It," p. 15.
46. Ibid.

47. *New York Times*, 22 October 1983, pp. 1, 13.
48. *New York Times*, 24 October 1983, p. D12.
49. Editorial, "The Medical/Industrial Complex," *Lancet* 2 (1973): 1380–81.
50. Barbara Culliton, "The Academic-Industrial Complex," *Science* 216 (1982): 960–62.
51. Ibid.
52. Levine, *Love Canal*, pp. 71–114.
53. Keith Schneider, "Reagan Gives Researchers Chills," *In These Times*, 27 April–3 May 1983, pp. 14, 15.
54. Ibid.
55. *New York Times*, 2 December 1982, p. B17.
56. Doll and Peto, "Causes of Cancer."
57. Nathan Karch and Marvin Schneiderman, "Explaining the Urban Factor in Lung Cancer Mortality," Report to the Natural Resources Defense Council (Washington, D.C., 1981).

3: Environmental Regulations

1. The case history is based on the following sources: Washington Heights Health Action Project, *Report on Conference on Hazardous Materials in the Metropolitan Region*, Columbia University, New York, N.Y., 13 June 1981, pp. 14–16; Nick Freudenberg and Sally Kohn, "The Washington Heights Health Action Project: A New Role for Social Service Workers," *Catalyst* 4, no. 1 (1982): 7–24; *New York Times*, 8 August 1980, pp. A1, B2; 9 August 1980, pp. 1, 24; 13 August 1980, p. B6; 20 September 1980, p. B3; 6 October 1980, p. B3; 2 September 1982, p. B3.
2. U.S. Congress, Senate, Commitee on Commerce, Science, and Transportation, *Hazardous Materials Transportation: A Review and Analysis of the Department of Transportation's Regulatory Program* (Washington, D.C.: Government Printing Office, 1979).
3. *New York Times*, 27 July 1982, p. B3; *New York Times*, 3 September 1982, p. B3.
4. Press release from Office of Mayor Edward I. Koch, 30 July 1982.
5. *New York Times*, 28 February 1984, p. A1.
6. Marion Edey, "Eco-politics and the League of Conservation Voters," in *The Environmental Handbook*, ed. Garrett De Bell (New York: Ballantine, 1970), p. 312.
7. Council on Environmental Quality, *Environmental Quality—10th Annual Report* (Washington, D.C.: Council on Environmental Quality, 1979); *New York Times*, 29 June 1983, p. A20.
8. A. Myrick Freeman III, *The Benefits of Air and Water Pollution Control* (Washington, D.C.: Council on Environmental Quality, 1979).
9. K. R. Mahaffey et al., "National Estimates of Blood Lead Levels: United States 1976–1980: Association with Selected Demographic and Socioeconomic Factors," *New England Journal of Medicine* 307 (1982): 573–79.
10. *New York Times*, 10 May 1983, p. A1.
11. *Atlantic Monthly*, June 1980, p. 7.
12. *Atlantic Monthly*, October 1980, p. 2.
13. *New York Times*, 26 August 1982, p. A27.
14. *Atlantic Monthly*, April 1980, p. 26.
15. Quoted in *New York Times*, 10 June 1983, p. B7.
16. Quoted in *New York Times*, 3 November 1981, p. A19.
17. *New York Times*, 2 June 1983, p. A19.

280   References

18. *In These Times,* 27 July–9 August 1983, pp. 4–5.
19. *New York Times,* 23 May 1983, p. A1.
20. Northwest Coalition Against Pesticides [NCAP] Staff, "The Saga of 2,4,5-T," *NCAP News* 3, no. 1 (1981–1982): 4–7.
21. Frederica Perera and Catherine Petito, "Formaldehyde: A Question of Cancer Policy?" *Science* 216 (1982): 1285–91.
22. John Walsh, "EPA and the Toxic Substances Law: Dealing with Uncertainty," *Science* 202 (1978): 598–602.
23. Interview with Susan Sladack, Assistant for Congress and Intergovernmental Affairs, Region I, EPA, Boston, July 1982.
24. Walsh, "EPA and the Toxic Substances Law."
25. Louis Harris, "Campaign 1982," *The Amicus Journal* 4, no. 1 (1982): 8–9.
26. *New York Times,* 15 April 1982, p. B13.
27. "A Clean Sweep?" *Environmental Action* 15, no. 1 (1983): 6.
28. Ibid.
29. James Ridgeway, *The Politics of Ecology* (New York: Dutton, 1971), p. 209.
30. *New York Times,* 19 November 1982, p. A18.
31. Odom Fanning, *Man and His Environment: Citizen Action* (New York: Harper & Row, 1975), pp. 1–24.
32. *New York Times,* 25 April 1982, p. 31.
33. NCAP Staff, "The Saga of 2,4,5-T."
34. Quoted in Daniel Berman, *Death on the Job* (New York: Monthly Review Press, 1978), pp. 33–34.
35. *Philadelphia Inquirer,* 25 September 1979.
36. *New York Times,* 27 October 1981, p. A16.
37. *New York Times,* 20 December 1983, p. A18.
38. Mark Green and Norman Waitzman, *Business War on the Law* (Washington, D.C.: Corporate Accountability Research Group, 1981).
39. "Waste Exports," *Exposure* 9 (1981): 1–2.
40. Richard Kazis and Richard Grossman, *Fear at Work* (New York: Pilgrim Press, 1982).
41. Jim Detjen, "PCBs in the Hudson River," in *Who's Poisoning America?* ed. Ralph Nader, Ronald Brownstein, and John Richard (San Francisco: Sierra Club Books, 1981), pp. 170–205.
42. *New York Times,* 12 June 1983, sect. 3, pp. 6–7.
43. Quoted in *New York Times,* 26 December 1982, p. 22.
44. Bradley Marten, "Regulation of the Transportation of Hazardous Materials: A Critique and a Proposal," *Harvard Environmental Law Review* 5 no. 2 (1981): 345–76.
45. Ibid.
46. *Los Angeles Times,* 3 September 1980.
47. Quoted in ibid.
48. *Los Angeles Times,* 1 September 1980.
49. *Los Angeles Times,* 22 July 1979.
50. "On the Trail of the Filthy Five," *Environmental Action* 13, no. 10 (1982): 26–27.
51. Quoted in *New York Times,* 22 April 1982, p. A25.
52. Quoted in *New York Times,* 9 July 1982, p. A22.
53. June Taylor, "Aspiring to Be Third Rate," *Amicus Journal* 3, no. 3 (1982): 18.
54. *New York Times,* 16 January 1983, p. 41.
55. *New York Times,* 23 January 1982, p. 1.
56. Taylor, "Aspiring to Be Third Rate," p. 17.
57. *New York Times,* 15 June 1983, p. A23.
58. Taylor, "Aspiring to Be Third Rate," p. 14.

59. Ibid., p. 15.
60. Quoted in *New York Times*, 28 April 1983, p. A15.
61. Quoted in ibid.
62. *New York Times*, 21 June 1981, p. 1.
63. *New York Times*, 7 April 1983, p. B1.
64. *New York Times*, 13 February 1983, p. E4.
65. Ibid.
66. *New York Times*, 10 February 1983, pp. A1, B10.
67. *New York Times*, 23 March 1983, p. A22.
68. Executive Order No. 12,291, 46 *Federal Register* 13, 193 (1981).
69. Kazis and Grossman, *Fear at Work*, p. 98.
70. Industrial Union Department, AFL-CIO v. American Petroleum Institute, 448 U.S. 607, 100 S.Ct. 2844, 48 U.S.C.W. 5022 (1980).
71. Victor Kimm, Arnold Kuzmack, and David Schnare, "Waterborne Carcinogens: A Regulator's View," in *The Scientific Basis of Health and Safety Regulation*, ed. Robert Crandall and Lester Lave (Washington, D.C.: Brookings Institute, 1981), p. 238.
72. David Noble, "Cost-Benefit Analysis: The Regulation of Business or Scientific Pornography?" *HealthPAC Bulletin* 11, no. 6 (1980): 7.
73. Luther Carter, "An Industry Study of TSCA: How to Achieve Credibility?" *Science* 203 (1979): 247–49.
74. Kimm, Kuzmack, and Schnare, "Waterborne Carcinogens," p. 248.
75. Leslie Boden, "Cost-Benefit Analysis: Caveat Emptor," *American Journal of Public Health* 69 (1979): 1200–11.
76. Kimm, Kuzmack, and Schnare, "Waterborne Carcinogens," p. 241.
77. *Federal Regulation and Regulatory Reform*, Report of the Subcommittee on Oversight and Investigation of the Committee on Interstate and Foreign Commerce, U.S. House of Representatives, 95th Congress, Second Session, October 1976.
78. Boden, "Cost-Benefit Analysis," p. 1211.
79. Noble, "Cost-Benefit Analysis," p. 11.
80. Quoted in ibid.
81. *New York Times*, 13 July 1983, pp. A1, A12; 6 November 1983, p. 31.
82. William Ruckleshaus, "How E.P.A. Faces the Arsenic Risk," *New York Times*, 23 July 1983, p. 22.
83. Quoted in *New York Times*, 14 July 1983, p. A12.

4: Arming Yourself for Battle

1. The case history is based on the following sources: *Staten Island Advance*, 22 July 1976, 30 July 1976, 6 August 1976, 5 May 1979; *Village Voice*, 28 June 1976; various flyers published by BLAST 1974–79; and interview with Edwina Cosgriff, November 1981.
2. *Staten Island Advance*, 26 May 1982, p. 1.
3. Quoted in BLAST, "The 'Cold' Facts of LNG," Mimeographed, n.d., p. 2.
4. G. Pascal Zachary, "Sloppy Design and Poor NRC Review Plague Diablo," *In These Times*, 16–22 December 1981, p. 7.
5. John Mitchell, "The Bedford Syndrome," *Massachusetts Audubon*, January 1980.
6. David Brook, *Stream Walk Manual* (Trenton, N.J.: New Jersey Public Interest Research Group, 1978).
7. *New York Times*, 28 July 1981, p. B4.
8. *Los Angeles Times*, 30 July 1979.

9. Nick Freudenberg and Sally Kohn, "The Washington Heights Health Action Project: A New Role for Social Service Workers," Catalyst 4, no. 1 (1982): 7–23.
10. Greater Newark Bay Coalition Against Toxic Wastes, "SCA-A company to be trusted with handling dangerous toxic materials?" undated flyer.
11. Francesca Lyman, "Locking Up Federal Files," Environmental Action 13, no. 2 (1981): 22–24.
12. Ibid.
13. Ibid.
14. Ibid.
15. New York Times, 16 August 1983.
16. Rachel Carson, Silent Spring (Boston: Houghton Mifflin, 1962).
17. Barry Commoner, The Closing Circle (New York: Alfred Knopf, 1971).
18. Samuel Epstein, The Politics of Cancer (Garden City, N.Y.: Anchor Books, 1979).
19. Samuel Epstein, Lester Brown, and Carl Pope, Hazardous Waste in America (San Francisco: Sierra Club Books, 1982).
20. Joseph Bell, "What Happened to My Baby?" McCalls, January 1980, p. 12.
21. Interview with Heather Baird Barney, research scientist for Massachusetts Fair Share, July 1982.
22. Ibid.
23. Lois Gibbs, "Lessons from Love Canal," speech at the Conference on Hazardous Materials in the Metropolitan Region, Columbia University, New York, N.Y., 13 June 1981.
24. Norton Kalishman, A. K. Hottle, and Bonnie Hill, "A Grassroots Concern about Herbicide Spraying Leads to Nationwide Ban," paper presented at American Public Health Association Annual Meeting, New York, N.Y., November 1979.
25. The Sun (Bridgehampton, N.Y.), 12 August 1982, p. 8.
26. Michael Brown, Laying Waste (New York: Pantheon, 1980).
27. Paul Brodeur, Expendable Americans (New York: Viking, 1973); The Zapping of America (New York: Norton, 1977).
28. Conservation Foundation, State of the Environment 1982 (Washington, D.C.: Conservation Foundation, 1982).
29. National Clean Air Coalition, Major Industrial Sources of Potential Toxic Air Pollutants (Washington, D.C.: National Clean Air Coalition, 1982).
30. Interview with Edwina Cosgriff.
31. Quoted in Flint Township–Swartz Creek News (Michigan), 18 March 1981, p. 3.
32. Interview with Helene Brathwaite, January 1982.

5: Building an Organization that Can Survive

1. The case history is based on the following sources: Rick Levine, "Cancer Town," New Times, 7 August 1978, pp. 21–32; New York Times, 4 April 1978, p. 69; 15 April 1978, p. 51; 2 May 1978, p. 2; 10 May 1978, p. 25; and interview with Carol Froelich and Louisa Nichols, December 1981.
2. Interview with Carol Froelich.
3. Ibid.
4. Quoted in New York Times, 15 April 1978, p. 51.
5. William Halperin et al., "Epidemiological Investigations of Clusters of Leukemia and Hodgkin's Disease in Rutherford, New Jersey," Journal of the Medical Society of New Jersey 77 (1980): 267–73.
6. Interview with Carol Froelich.

7. Quoted in Levine, "Cancer Town," p. 32.
8. Interview with Edwina Cosgriff, November 1981.
9. Bob Biagi, *Working Together: A Manual for Helping Groups Work More Effectively* (Amherst: University of Massachusetts [Citizen Involvement Training Project], 1978), p. 112.
10. Rhoda Linton and Michele Whitman, "With Mourning, Rage, Empowerment, and Defiance," *Socialist Review* 12, nos. 3–4 (1982): 11–36.
11. Greg Speeter, *Power: A Repossession Manual* (Amherst: University of Massachusetts [Citizen Involvement Training Project], 1978), p. 59.
12. Interview with Edwina Cosgriff.
13. Adeline Levine, *Love Canal: Science, Politics, and People* (Lexington, Mass.: Lexington Books, 1982), pp. 199–201.
14. Ibid., p. 203.
15. Interview with Edwina Cosgriff.
16. Nick Freudenberg and Sally Kohn, "The Washington Heights Health Action Project: A New Role for Social Service Workers in Community Organizing," *Catalyst* 4, no. 1 (1982): 7–24.
17. David Weinberg, "Breakdown, Love Canal's Walking Wounded," *Village Voice*, 15–25 September 1981, p. 11.
18. Quoted in *New York Times*, 23 August 1981, p. 61.
19. Interview with Helene Brathwaite, January 1982.

6: **Strategies for Action: Community Education**

1. The case history is based on the following sources: Caron Chess, *Winning the Right to Know: A Handbook for Toxics Activists* (Philadelphia: Delaware Valley Toxics Coalition, 1983); "Interview with Caron Chess and Jim Moran," *Exposure*, May 1981, pp. 4–5, 8; *New York Times*, 23 January 1981, p. A14; Caron Chess, "Working in the Legislative Arena," speech at the Conference on Hazardous Materials in the Metropolitan Region, Columbia University, New York, N.Y., 13 June 1981.
2. Quoted in *New York Times*, 23 January 1981, p. A14.
3. Quoted in ibid.
4. Chess, "Working in the Legislative Arena."
5. "Interview with Chess and Moran," p. 4.
6. Ibid., p. 5.
7. *New York Times*, 10 November 1983, p. B12.
8. Testimony in "Song of the Canary," film produced by Josh Hanig and Dave Davis, distributed by New Day Films, Franklin Lakes, N.J.
9. Fred Wilcox, "Balloon Releases," in *Grassroots: An Antinuclear Source Book*, ed. Fred Wilcox (Trumansburg, N.Y.: Crossing Press, 1980), pp. 98–100.
10. Interview with Edwina Cosgriff, November 1981.
11. Committee to Defend Reproductive Rights, *The Media Book* (San Francisco, 1981), p. 10.
12. Lois Gibbs, "Lessons from Love Canal," speech at Conference on Hazardous Materials in the Metropolitan Region, Columbia University, New York, N.Y., 13 June 1981.
13. Interview with David Wilson, October 1981.
14. Cathy Wolff and Steven Hilgartner, "Working with the Media—A Case History," in *We Interrupt This Program . . . A Citizen's Guide to the Media for Social Change*, ed. Robbie Gordon (Amherst, Mass.: University of Massachusetts [Citizen Involvement Training Project], 1978), p. 37.
15. Interview with Carol Froelich, December 1981.

16. Lois Gibbs, "Lessons from Love Canal."
17. Anna Gyorgy, No Nukes (Boston: South End Press, 1979), p. 404.
18. Wolff and Hilgartner, "Working with the Media," p. 40.
19. Todd Gitlin, The Whole World Is Watching (Berkeley: University of California Press, 1980).
20. Interview with Helene Brathwaite, January 1982.
21. Ben H. Bagdikian, The Media Monopoly (Boston: Beacon Press, 1983), pp. 8–9.
22. Ibid., pp. 24–25.
23. Quoted in New York Times, 18 December 1983, p. F4.
24. Gibbs, "Lessons from Love Canal."
25. Quoted in New York Times, 26 May 1983, p. A14.

7: Strategies for Action: Legal and Legislative

1. The case history is based on the following sources: Harvey Wasserman, "We All Live Downwind," Environmental Action 14, no. 8 (1983): 16–19; New York Times, 5 August 1982, p. B13; 15 August 1982, p. 18; 25 August 1982, p. A12; 14 September 1982, p. A16; 19 September 1982, p. 31; 21 September 1982, p. A13; 21 September 1982, p. A12; 27 September 1982, p. B5; 3 October 1982, p. 73; 10 October 1982, p. A18; 17 November 1982, p. A16; 18 December 1982, p. 12.
2. Quoted in New York Times, 15 August 1982.
3. Quoted in New York Times, 5 August 1982.
4. New York Times, 27 November 1983.
5. Quoted in New York Times, 14 September 1982.
6. New York Times, 21 September 1982, p. A12.
7. Ibid.
8. New York Times, 10 October 1982.
9. New York Times, 27 September 1982.
10. Ibid.
11. New York Times, 17 November 1982.
12. New York Times, 11 May 1984, pp. A1, A17.
13. Interview with Jacinta Fernandes and Gloria Davis, March 1982.
14. The case history is based on the following sources: San Francisco Examiner, 12 April 1979; Los Angeles Times, 19 June 1979.
15. Quoted in San Francisco Examiner, 12 April 1979.
16. Quoted in ibid.
17. Quoted in Los Angeles Times, 19 June 1979.
18. Quoted in ibid.
19. Quoted in ibid.
20. The case history is based on the following source: Findlay d'Arbois, "Industry Dollars Inundate Herbicide Debate," NCAP News (Northwest Coalition for Alternatives to Pesticides) 2, no. 3 (1981): 30–33.
21. Quoted in ibid.
22. Quoted in ibid.
23. Frank Graham Jr., Since Silent Spring (Boston: Houghton Mifflin, 1970).
24. Quoted in Stephen Fox, John Muir and His Legacy (Boston: Little, Brown & Company, 1981), p. 304.
25. New York Times, 9 September 1981, p. A12.
26. New York Times, 19 December 1982, p. 79.
27. Office of Water and Waste, Siting of Hazardous Waste Management Facilities (Washington, D.C.: Environmental Protection Agency, EP-530/SW-809, 1979), p. 14.

28. *New York Times*, 28 November 1982, p. 64.
29. Scott Reed, "Contemplating a Lawsuit?" *NRAG Papers* (Northern Rockies Action Group) 4, no. 1 (1981): 14.
30. Interview with Nell Grantham, October 1981.
31. David Wilson, speech at Chemicals in the Environment Conference, Scarritt College, Nashville, Tenn., 27 September 1981.
32. *New York Times*, 1 May 1982, p. 30.
33. *New York Times*, 27 August 1982, p. D4.
34. Mike Males, "Be It Enacted by the People: A Citizens' Guide to Initiatives," *NRAG Papers* 4, no. 2 (1981): 27.
35. Richard Pollock, "California's Kern County Votes Down Nukes 2-1," in *Grassroots*, ed. Fred Wilson (Trumansburg, N.Y.: Crossing Press, 1980), pp. 90–91.
36. William Reynolds, "Communities Block Shipments," in *Grassroots*, pp. 66–67.
37. *New York Times*, 7 November 1982, p. 36.
38. Quoted in Ginny Thomas and Bill Brooks, " 'Buddy, We're Home': Halting the Heard County Landfill," *Southern Exposure* 9, no. 3 (1981): 40.
39. Elizabeth Davenport, "A Letter to Our Members," *Environmental Action* 14, no. 5 (1982/1983): 21.
40. Males, "Be It Enacted," p. 27.
41. "Interview with Caron Chess and Jim Moran," *Exposure*, May 1981, p. 5.
42. Interview with Edwina Cosgriff, November 1981.
43. Males, "Be It Enacted," p. 27.
44. Steve Lydenberg, *Bankrolling Ballots* (New York: Council on Economic Priorities, 1979), pp. 1–9.
45. Interview with Carol Froelich, December 1981.
46. Mike Miller, "Community Organization Vision and the Electoral Tactic," *Socialist Review* 12, nos. 3–4 (1982): 188.
47. Ibid.
48. Thomas and Brooks, " 'Buddy, We're Home,' " pp. 38–41.
49. Interview with Gregor MacGregor, July 1982.
50. Reed, "Contemplating a Lawsuit?" p. 30.

8: Strategies for Action: Community Organizing

1. The case history is based on the following source: Association of Community Organizations for Reform Now (ACORN), Memorandum from Glen Runkle, Forth Worth ACORN, to Larry Ginsberg, ACORN Research Staff, Mimeographed, n.d.
2. The case history is based on the following sources: "Concerned Citizens Back Up PCB Dump Trucks," *Federation for Progress Newsletter* 2, no. 2 (1982): 1–2; *In These Times*, 29 September–5 October 1982, pp. 5–6; *New York Times*, 16 September 1982, p. A18; 21 September 1982, p. A10; 25 October 1982, p. 31; and speeches by Kenneth Ferruccio and the Reverend Leon White, Forum on Toxic Dumping in Warren County, North Carolina, District 1199, New York, N.Y., 3 December 1982.
3. Michael Brown, *Laying Waste: The Poisoning of America by Toxic Chemicals* (New York: Pantheon, 1979), p. 244.
4. Ibid., p. 245.
5. *New York Times*, 25 September 1982, p. 17.
6. *In These Times*, 29 September–5 October 1982, p. 5.
7. Quoted in ibid.
8. *New York Times*, 21 September 1982, p. A10; 28 September 1982, p. A16.

9. Ferruccio, Speech at Forum.
10. *New York Times*, 10 October 1982, p. 31.
11. "Concerned Citizens Back Up PCB Dump Trucks," p. 2.
12. Heather Booth, *Direct Action Organizing* (Chicago: Midwest Academy, 1977), p. 3.
13. Massachusetts Fair Share Health and Safety Project, *Neighborhood Health and Safety Manual* (Boston: Massachusetts Fair Share, 1983), p. ii.
14. Ibid., p. i.
15. *Sun* (Lowell, Mass.), 10 August 1981.
16. "Mass Fair Share," *Exposure* 9 (1981): 3.
17. *Bulletin* (Philadelphia), 23 August 1981, p. 1.
18. Barry M. Casper and Paul David Wellstone, *Powerline* (Amherst: University of Massachusetts Press, 1981).
19. Ibid., p. 136.
20. Ibid., p. 207.
21. Ibid., pp. 215–16.
22. Ibid., p. 220.
23. Ibid., p. 279.
24. Ibid., p. 288.
25. Ibid., pp. 289–90.
26. *New York Times*, 21 May 1980.
27. Adeline Levine, *Love Canal: Science, Politics and People* (Lexington, Mass.: Lexington Books, 1982), p. 149.
28. Interview with Nell Grantham, October 1981.
29. Massachusetts Fair Share Health and Safety Project, *Neighborhood Health and Safety Manual*, p. 200.2.
30. Frances Fox Piven and Richard A. Cloward, *Poor People's Movements* (New York: Vintage Books, 1977).
31. Casper and Wellstone, *Powerline*, p. 137.
32. Ibid., p. 196.
33. "Taking It to the Streets: An Interview with Wade Rathke, Chief Organizer of ACORN," *Actions and Campaigns* (New Orleans: The Institute, 1979), p. 29.
34. Casper and Wellstone, *Powerline*, p. 203.

9: Coalition Building: Issues and Problems

1. The case history is based on the following sources: Cathy Lerza, "Environmental Issues Reach the Bargaining Table," *Environmental Action* 4, no. 20 (1973): 3–6; Daniel M. Berman, *Death on the Job* (New York: Monthly Review, 1978), pp. 52, 122, 123; and Richard Kazis and Richard Grossman, *Fear at Work* (New York: Pilgrim Press, 1982), pp. 235–38.
2. Lerza, "Environmental Issues," p. 4.
3. Ibid., p. 3.
4. Barry Commoner, "The Promise and Perils of Petrochemicals," in *The Big Business Reader*, ed. Mark Green and Robert Massie, Jr. (New York: Pilgrim Press, 1980), p. 349.
5. *In These Times*, 8–14 April 1981, p. 2.
6. *New York Times*, 2 May 1973.
7. Quoted in Kazis and Grossman, *Fear At Work*, p. 247.
8. Reported in Harvey Wasserman, "Unionizing Ecotopia," *Mother Jones*, June 1978, p. 31.
9. Interview with Edwina Cosgriff, November 1981.
10. Quoted in Wasserman, "Unionizing Ecotopia," p. 33.

11. *In These Times*, 11–17 November 1981, p. 2.
12. *In These Times*, 8–14 April 1981, p. 5.
13. Lerza, "Environmental Issues," p. 6.
14. Barry Commoner, *Alliance for Survival* (New York: United Electrical Workers of America, 1972), p. 5.
15. Kazis and Grossman, *Fear at Work*, p. 253.
16. Deborah Baldwin, "Surf and Turf," *Environmental Action* 11, no. 3 (1979): 11–16.
17. Interview with George McDevitt, March 1982.
18. Ibid.
19. Quoted in Kazis and Grossman, *Fear at Work*, p. 58.
20. Liv Smith, "Labor and the No-Nukes Movement," *Socialist Review* 10, no. 6 (1980): 137.
21. Quoted in Wasserman, "Unionizing Ecotopia," p. 34.
22. Interview with George McDevitt.
23. Cynthia Crowner, "Community, Workers Indict Oil Industry," *Union Wage* (Women's Alliance to Gain Equality, San Francisco) 66 (July–August 1981): 1–2.
24. Peter Medoff, "A Talk with Tony Mazzocchi," *Health/PAC Bulletin* 13, no. 2 (1982): 9–10.
25. Smith, "Labor and the No-Nukes Movement," p. 148.
26. Quoted in Kazis and Grossman, *Fear at Work*, pp. 253–54.
27. T. F. Fleming, "Job Creating Potential of Federal Spending," *Monthly Labor Review*, November 1975.
28. Quoted in *New York Times*, 8 February 1982, p. B2.
29. Quoted in Wasserman, "Unionizing Ecotopia," p. 37.
30. The case history is based on the following sources: *New York Times*, 15 November 1978, sect. 2, p. 2; 21 November 1978, sect. 2, p. 2; 6 December 1978, sect. 2, p. 2; and interview with Helene Brathwaite, January 1982.
31. *New York Times*, 21 November 1978, sect. 2, p. 2.
32. Interview with Helene Brathwaite.
33. Douglas Strong, "The Sierra Club—A History—Part 2. Conservation." *Sierra Club Bulletin*, November–December 1977, p. 16.
34. David Kotelchuck, "Occupational Injuries and Illnesses among Black Workers," *Health/PAC Bulletin* 81/82 (1979): 33–34.
35. A. J. McMichael et al., "Mortality among Rubber Workers: Relationship to Specific Jobs," *Journal of Occupational Medicine* 18, no. 2 (1976): 178–85; J. W. Lloyd et al., "Long-term Mortality Study of Steelworkers, IV: Mortality by Work Area," *Journal of Occupational Medicine* 12 (1970): 151–57; A. Blair, P. Decoufle, and D. Grauman, "Causes of Death among Laundry and Dry Cleaning Workers," *American Journal of Public Health* 69 (1979): 508–11; W. J. Blot et al., "Lung and Laryngeal Cancers in Relation to Shipyard Employment in Coastal Virginia," *Journal of the National Cancer Institute* 65 (1980): 571–75.
36. Office of Health Resources Opportunity, *Health Status of Minorities and Low-Income Groups*, DHEW Publication No. (HRA) 79-627 (Washington, D.C.: Government Printing Office, 1979), p. 8.
37. Kotelchuck, "Occupational Illnesses and Injuries among Black Workers," p. 34.
38. Susan L. Pollack, *The Hired Farm Working Force of 1979*, USDA/ERA, Agricultural Economic Report #473 (Washington, D.C.: Government Printing Office, August 1981), p. 24.
39. F. W. Kutz, A. R. Yobs, and S. C. Strassman, "Racial Stratification of Or-

ganochlorine Insecticide Residues in Human Adipose Tissues," *Journal of Occupational Medicine* 19 (1977): 6219–29.

40. A. Joseph D'Ercole et al., "Insecticide Exposure of Mothers and Newborns in a Rural Agricultural Area," *Pediatrics* 57 (1976): 869–74.

41. Franz Leichter, "The Return of the Sweatshop," Report to the New York State Senate, 26 February 1981.

42. Ellen Hall, *Inner City Health in America* (Washington, D.C.: Urban Environmental Foundation, 1979); *New York Times*, 2 September 1982, p. A13; and Edward Radford and Terence Drizd, "Blood Carbon Monoxide Levels in Persons 3–74 Years of Age: United States, 1976–80," *Advance Data* 76 (1982): 8.

43. Eleanor J. MacDonald, "Demographic Variation in Cancer in Relation to Industrial and Environmental Influence," *Environmental Health Perspectives* 17 (1976): 153–66.

44. Madonna Thunder Hawk and Margie Bowker, "Listening to Native American Women," *Heresies* 13 (1981): 21.

45. Barry Commoner, *The Closing Circle* (New York: Alfred Knopf, 1971), p. 205.

46. Stephen Fox, *John Muir and His Legacy* (Boston: Little Brown, 1981), p. 324.

47. Anna Gyorgy, *No Nukes* (Boston: South End Press, 1979), p. 389.

48. Adeline Levine, *Love Canal: Science, Politics, and People* (Lexington, Mass.: Lexington Books, 1982), p. 198.

49. Fox, *John Muir and His Legacy*, p. 324.

50. Interview with Margaret Williams, October 1981.

51. Interview with Jacinta Fernandes, March 1982.

52. Lois Gibbs, "Lessons from Love Canal," speech at the Conference on Hazardous Materials in the Metropolitan Region, Columbia University, New York, N.Y., 13 June 1981.

53. The case history is based on a response to written survey by Women Opposed to Nuclear Technology, 1981.

54. The case history is based on the following sources: Rhoda Linton and Michelle Whitman, "With Mourning, Rage, Empowerment and Defiance: The 1981 Women's Pentagon Action," *Socialist Review* 12, nos. 3–4 (1982): 11–36; "Mobilizing Emotions Organizing the Women's Pentagon Action, Interview with Donna Warnock," *Socialist Review* 12, nos. 3–4 (1982): 37–47.

55. Linton and Whitman, "With Mourning, Rage, Empowerment and Defiance," p. 11.

56. "Mobilizing Emotions," p. 38.

57. Linton and Whitman, "With Mourning, Rage, Empowerment and Defiance," p. 18.

58. "Mobilizing Emotions," p. 43.

59. Zillah R. Eisenstein, ed., *Capitalist Patriarchy and the Case for Socialist Feminism* (New York: Monthly Review, 1979), p. 17.

60. Susan Koen and Nina Swaim, *Ain't No Where We Can Run* (Norwich, Vt.: Women Against Nuclear Development, 1980).

61. Interview with Helene Brathwaite.

62. Wendy Chavkin, ed., *Double Exposure: Women's Health Hazards on the Job and at Home* (New York: Monthly Review, 1984), pp. 157–59.

63. The case history is based on the following sources: Robert Aldridge, "The Underground Trigger Finger," *The Nation*, 16 June 1979; various pamphlets produced by Stop Project ELF; and response to written survey by Stop Project ELF, 1982.

64. *In These Times*, 12-18 January 1983, p. 6.

65. U.S. General Accounting Office, *The Navy's Strategic Communication System—Need for Management Attention and Decisionmaking*, PSAD-79-48A (Washington, D.C.: Government Printing Office, 1979).

66. Jennifer Speicher and John Stauber, "Letter to Prospective Members," Stop Project ELF, Summer 1981.
67. Stop Project ELF, "Guide to Stop Project ELF Inc.'s Response to and Review of Navy Report," December 1981, pp. 2 and 4.
68. *In These Times*, 12–18 January 1983, p. 7.
69. *New York Times*, 8 September 1982, p. A14.
70. Gail Robinson, "The War Hits Home," *Environmental Action* 13, no. 3 (1981): 18.
71. Quoted in ibid.
72. Quoted in Michael Kazin, "Politics and the New Peace Movement," *Socialist Review*, 13, no. 1 (1983): 111.
73. Quoted in *New York Times*, 7 March 1982, p. A12.
74. Roger Burbach and Patricia Flynn, *Agribusiness in the Americas* (New York: Monthly Review Press, 1980), pp. 117–18, 159.
75. Chris Jenkins, "DBCP Runaway Hazard," *NACLA Report* 13, no. 2 (1979): 43–45.
76. Paul Horowitz, "Puerto Rico's Pharmaceutical Fix," *NACLA Report*, 15, no. 2 (1981): 22–26; *New York Times*, 24 July 1983, pp. 1, 18.
77. "Exports-Products," *Exposure* 9 (1981): 1.
78. Ibid.
79. "Exports-Waste," *Exposure* 9 (1981): 1–2.
80. "Pesticide Update," *Food First News* 14 (Spring 1983): 3.
81. Ibid.
82. Kim Gazella, "Chemicals Abroad," *Environmental Action* 14, no. 4 (1982): 4–5.

10: Toward a National Environmental Movement

1. The case history is based on the following sources: Margaret A. McKean, *Environmental Protest and Citizen Politics in Japan* (Berkeley: University of California Press, 1981); Paul M. Sweezy, "Japan in Perspective," *Monthly Review* 31, no. 9 (1980): 1–14; Herbert Bix, "Japan's New Vulnerability," *Monthly Review* 34, no. 7 (1982): 10–17; John Junkerman, "The Japanese Model," *Progressive*, May 1983, pp. 21–27; and Anna Gyorgy, *No Nukes* (Boston: South End Press, 1979), pp. 361–66.
2. McKean, *Environmental Protest and Citizen Politics in Japan*, pp. 51–59.
3. Ibid., pp. 59–61.
4. Ibid., pp. 45–50.
5. Ibid., pp. 88–97.
6. Gyorgy, *No Nukes*, p. 361.
7. Ibid., pp. 363–65.
8. McKean, *Environmental Protest and Citizen Politics in Japan*, pp. 223–68.
9. Ibid., p. 72.
10. The case history is based on the following sources: John Ely, "The Greens: Ecology and the Promise of Radical Democracy," *Radical America* 17, no. 1 (1983): 23–34; Carl Boggs, "The Greens, Anti-Militarism and the Global Crisis," *Radical America* 17, no. 1 (1983): 7–20; Andrei Markovits, "West Germany's Political Future: The 1983 Bundestag Elections," *Socialist Review* 13, no. 4 (1983): 67–98; George Katsiaficas, "The Extraparliamentary Left in Europe," *Monthly Review* 34, no. 4 (1982): 31–45; *In These Times*, 23–29 March 1983, pp. 3, 10; Frederick Painton, "Protest by the 'New Class,'" *Time*, 28 February 1983, pp. 30–31; James Markham, "Germany's Volatile Greens," *New York Times Magazine*, 13 February 1983, pp. 37, 70–77; Gyorgy, *No Nukes*, pp. 345–54.

11. Gyorgy, *No Nukes*, pp. 350–51.
12. Katsiaficas, "The Extraparliamentary Left," p. 37.
13. *In These Times*, 23–29 March 1983, p. 3.
14. Ely, "The Greens," p. 27.
15. Quoted in Markham, "Germany's Volatile Greens," p. 74.
16. Quoted in Ely, "The Greens," p. 27.
17. *In These Times*, 23–29 March 1983, p. 10.
18. Quoted in Gyorgy, *No Nukes*, p. 353.
19. Ely, "The Greens," p. 33.
20. Markham, "Germany's Volatile Greens," p. 72.
21. Cited in Anne Jackson and Angus Wright, "Nature's Banner," *Progressive*, October 1981, p. 26.
22. *New York Times*, 15 October 1983, p. 6.
23. U.S. Office of Water and Waste, *Siting of Hazardous Waste Management Facilities*, EP 530-SW-809 (Washington, D.C.: EPA, 1979), p. 25.
24. Stephen Fox, *John Muir and His Legacy* (Boston: Little Brown, 1981), p. 315.
25. Carl Pope, "Being Right Is Not Enough," *Environmental Action* 11, no. 10 (1980): 24–25.
26. *New York Times*, 7 November 1982, p. 36.
27. *New York Times*, 7 February 1982, p. 36.
28. Fox, *Muir and His Legacy*, p. 196.
29. National Wildlife Federation, *The Toxic Substances Dilemma* (Washington, D.C.: Government Printing Office, 1980), p. i.
30. Quoted in *New York Times*, 19 April 1981, p. 1.
31. Environmental Action, *Annual Report*, Washington, D.C., 1982.
32. Ibid., pp. 3–4.
33. Ibid., p. 4.
34. Ibid., p. 1.
35. Fox, *Muir and His Legacy*, pp. 328–29.
36. *New York Times*, 23 August 1982, p. A13.
37. *New York Times*, 19 September 1982, p. 34.
38. *New York Times*, 7 November 1982, p. 36.
39. Quoted in ibid.
40. "Environmentalists: More of a Political Force," *Business Week*, 24 January 1983, pp. 85–86.
41. Ibid.
42. Quoted in Frances Gendlin, "Mike McCloskey, Taking Stock, Looking Forward," *Sierra*, January/February 1983, p. 127.
43. David Foreman, "Earth First," *Progressive*, October 1981, pp. 39–42.
44. Ibid., p. 39.
45. Thomas Lovejoy, "Why Pupfish Matter," *New York Times*, 6 August 1982, p. A15.
46. See, for example, Marshall Goldman, *The Spoils of Progress* (Cambridge, Mass.: MIT Press, 1972); Boris Komarov, *The Destruction of Nature in the Soviet Union* (White Plains, N.Y.: Sharpe, 1980); *New York Times*, 31 January 1982, p. 9; Philip Pryde, "The 'Decade of the Environment' in the U.S.S.R.," *Science* 220 (1983): 274–79; I. Novik, *Society and the Environment: A Soviet View* (Moscow: Progress, 1982); and David Schwartzman, "What's New in the Noosphere?" *Environmental Action* 14, no. 6 (1983): 25–27.
47. In "Troubled Waters," a film produced by Meg Switzgable, 1982.

RESOURCES FOR ACTION

Books on Environmental Health Hazards

Brown, Michael. *Laying Waste: The Poisoning of America by Toxic Chemicals.* New York: Pantheon, 1980.

Carson, Rachel. *Silent Spring.* Boston: Houghton Mifflin, 1962.

Chavkin, Wendy, ed. *Double Exposure: Women's Health Hazards at Home and on the Job.* New York: Monthly Review Press, 1984.

Commoner, Barry. *The Closing Circle.* New York: Bantam, 1972.

Epstein, Samuel; Brown, Lester; and Pope, Carl. *Hazardous Waste in America.* San Francisco: Sierra Club, 1982.

Epstein, Samuel. *The Politics of Cancer.* Rev. ed., Garden City, N.Y.: Anchor Press, 1979.

Gofman, John. *Radiation and Human Health.* San Francisco: Sierra Club, 1982.

Gyorgy, Anna. *No Nukes.* Boston: South End Press, 1979.

Kazis, Richard, and Grossman, Richard L. *Fear at Work: Job Blackmail, Labor, and the Environment.* New York: Pilgrim Press, 1982.

Levine, Adeline. *Love Canal: Science, Politics, and People.* Lexington, MA: Lexington Books, 1982.

Nader, Ralph; Brownstein, Ronald; and Richard, John, eds. *Who's Poisoning America? Corporate Polluters and Their Victims in the Chemical Age.* San Francisco: Sierra Club, 1981.

Norwood, Christopher. *At Highest Risk: Environmental Hazards to Young and Unborn Children.* New York: McGraw-Hill, 1980.

Powledge, Fred. *Water: The Nature, Uses, and Future of Our Most Precious and Abused Resource.* New York: Farrar, Straus, and Giroux, 1982.

Stellman, Jeanne M., and Daum, Susan M. *Work Is Dangerous to Your Health.* New York: Vintage, 1973.

Van Strum, Carol. *A Bitter Fog: Pesticides and Human Rights.* San Francisco: Sierra Club, 1983.

Books on Community Organizing

Dale, Duane, and Mitiguy, Nancy. *Planning for a Change.* Citizen Involvement Training Project, Division of Continuing Education, University of Massachusetts, Amherst, MA 01003, 1978.

Flanagan, Joan. *The Grass Roots Fundraising Book.* Chicago: Swallow, 1977.

Kahn, Si. *How People Get Power.* New York: McGraw Hill, 1970.

————. *Organizing: A Guidebook for Grassroots Leaders.* New York: McGraw-Hill, 1982.

Lindberg, Mark. *Up With the Ranks: How Community Organizers Develop Community Leaders.* New England Training Center for Community Organizers, 23 Promenade St., Providence, RI 02908.

Shellow, Jill R. *The Grantseekers Guide.* Chicago: National Network of Grantmakers, 919 N. Michigan Ave., Chicago, Illinois 60611, 1981.

Specter, Greg. *Power: A Repossession Manual.* Citizen Involvement Training Project, Amherst MA, 1978.
Yale, David. *The Publicity Handbook.* New York: Bantam, 1982.

Manuals for Environmental Activists

Blanc, Paul. *Stop Environmental Cancer: A Citizen's Guide to Organizing.* Campaign for Economic Democracy, 409 Santa Monica Blvd., Santa Monica, CA 90401, 1980.
Chess, Caron. *Winning the Right to Know: A Handbook for Toxic Activists.* Delaware Valley Toxics Coalition, 1315 Walnut Street, Suite 1632, Philadelphia, PA 19107, 1983.
Citizen's Clearinghouse for Hazardous Waste. *Leadership Handbook.* CCHW, PO Box 926, Arlington, VA 22216, 1983.
Environmental Defense Fund. *Dumpsite Cleanups: A Citizen's Guide to the Superfund Program.* EDF, 1525 18th St. NW, Washington, D.C. 20036, 1983.
Koen, Susan, and Swaim, Nina. *Ain't No Where We Can Run: Handbook for Women on the Nuclear Mentality.* Women Against Nuclear Development, Box 421, Norwich, VT 05055, 1980.
League of Women Voters. *Protecting the California Environment: A Citizen's Guide.* LWV, 942 Market Street, Suite 505, San Francisco, CA 94102, 1980.
Massachusetts Fair Share Health and Safety Project. *Neighborhood Health and Safety Survival Manual.* Mass Fair Share, 304 Boylston Street, Boston, MA 02116, 1983.
Merrifield, Juliet. *We're Tired of Being Guinea Pigs!* Highlander Research and Education Center, Route 3, Box 370, New Market, TN 37820, 1980.
National Wildlife Federation. *The Toxic Substances Dilemma: A Plan for Citizen Action.* Washington, D.C.: U.S. Government Printing Office, 1980.
Sierra Club. *Training Materials on Toxic Substances: Tools for Effective Action,* 2 vols. San Francisco: Sierra Club, 1981.
Wilcox, Fred, ed. *Grass Roots: An Anti-Nuke Source Book.* Trumansburg, NY: Crossing Press, 1980.

Groups

The following is a list of national and regional organizations that can assist environmental activists and concerned residents. Unless otherwise noted, the group is in Washington, D.C.

ACORN (Association of Community Organizations for Reform Now)
413 8 St., SE 20003
Organizing: community improvement, utilities, taxes, health care, toxics.
(202) 547-9292

Center for Science in the Public Interest
1755 S St., NW 20009
Research and education: food, nutrition, health.
(202) 332-9110

Citizen/Labor Energy Coalition (C/LEC)
1300 Connecticut Ave., NW
Suite 401 20036
Research, education and lobbying: taxes on oil companies, solar and conservation, utility rate reform, natural gas deregulation.
(202) 857-5153

Citizens Clearinghouse for Hazardous Waste
P.O. Box 926, Arlington, VA 22216
Research and organizing: toxics, asbestos in schools.
(703) 276-7070

Citizens for a Better Environment
59 E. Van Buren, Suite 1600
Chicago, Ill. 60605
Research, education and litigation:
toxic substances, sludge management,
energy, air and water quality.
(312) 939-1530

Clean Water Action Project
733 15th St., NW, Suite 1110 20005
Lobbying: water quality.
(202) 638-1196

Conference on Alternative State and
Local Policies
200 Florida Ave., NW 20009
Research and education: cooperatives,
economic development, housing
finance, campaign finance reform,
women and the economy, energy conservation, toxics.
(202) 387-6030

Conservation Foundation
1717 Massachusetts Ave., NW 20036
Research and communications: land
use, energy conservation, air and water
quality, environmental mediation.
(202) 797-4300

Environmental Action Foundation
Dupont Circle Building, Suite 724
    20036
Research, education and organizing:
energy/nuclear economics, utility rates
and regulation, public power, alternatives to power plant construction,
solid and hazardous wastes, toxic substances.
(202) 659-9682.

Environmental Action Inc.
1346 Connecticut Ave., NW
Suite 731 20036
Education and lobbying: air quality,
deposit legislation, nuclear energy,
solid waste, toxic substances, political
action, arms control.
(202) 833-1845

Environmental Defense Fund
1525 18th St., NW 20036
Research, litigation and lobbying:
drinking water, toxics, pesticides and
wildlife.
(202) 387-3500

Environmental Law Institute
1346 Connecticut Ave., NW,
Suite 600 20036
Research, education and litigation: toxics victim compensation.
(202) 452-9600

Environmental Policy Center
317 Pennsylvania Ave., SE 20003
Research, education and lobbying: water, synthetic fuels, toxics, pesticides,
rural development, public lands, surface mining, energy conservation.
(202) 547-5330

Environmentalists for Full Employ-
    ment
1536 16th St., NW 20036
Research and education: jobs potential
of environmental protection, labor and
safe energy.
(202) 347-5590

Friends of the Earth
1045 Sansome St.
San Francisco, CA 94111
Lobbying: air quality, pesticides, nuclear energy, water policy, wildlife, energy, public lands.
(415) 433-7373

Greenpeace
2007 R St., NW 20009
Lobbying, organizing and direct action:
whales, nuclear, toxics.
(202) 462-1177

Health Research Group
2000 P St., NW, Suite 708 20036
Research, publications, education: toxic substances, health care and delivery,
occupational safety, food and drug
legislation.
(202) 872-0320

Highlander Center
Rt. 3, Box 370
New Market, Tenn. 37820
Research, education and training: land
use, strip mining, corporate responsibility, utilities, toxics.
(615) 933-3443

League of Conservation Voters
317 Pennsylvania Ave., SE 20003
Political action and organizing: evaluation of environmental records of federal officials.
(202) 546-5246

League of Women Voters of the U.S.
1730 M St., NW 20036
Lobbying and education: air and water
quality, solid waste, toxics.
(202) 429-1965

Midwest Academy
600 W. Fullerton, Chicago, Ill. 60614
Training and organizing: leadership
development, fundraising skills.
(312) 975-3670

Mobilization for Survival
3601 Locust Walk,
Philadelphia, Penn. 19104
Organizing: nuclear weapons, the arms
race, nuclear power, human needs.
(215) 386-4875

National Association of
Farm Worker Organizations
1316 10th St., NW 20001
Research and education: pesticides
and other occupational health hazards.
(202) 328-9777

National Association of Neighbor-
    hoods
1651 Fuller St., NW 20009
Research, education and lobbying:
neighborhood organizing, housing, environmental quality.
(202) 332-7766

National Audubon Society
645 Pennsylvania Ave., SE 20003
Research and lobbying: wildlife, wilderness, public lands, endangered
species, water policy, energy.
(202) 547-9009

National Clean Air Coalition
530 7th St., SE 20003
Lobbying: air quality, acid rain.
(202) 543-8200

National Wildlife Federation
1412 16th St., NW 20036
Research and education: wildlife, wilderness, environmental quality.
(202) 797-6800

National Women's Health Network
224 7th St., SE 20003
Research and education: reproductive
rights of workers, environmental and
occupational health, national health
care.
(202) 543-9222

Natural Resources Defense Council
1725 Eye St., NW, Suite 600 20006
Research, organizing and litigation:
water and air quality, land use, energy,
forestry, agriculture, international environmental issues, nuclear nonprolif-
eration.
(202) 223-8210

New York Committee for Occupational
    Safety and Health
32 Union Square E., New York, N.Y.
    10003
Education and organizing: occupational hazards (send for list of other
COSH groups around the country).
(212) 674-1595

Northern Plains Resource Council
419 Stapleton Bldg.
Billings, Mont. 59101
Education, organizing and lobbying:
energy, water, strip mining, public
lands.
(406) 248-1154

Northwest Coalition for
Alternatives to Pesticides
Box 375, Eugene, Ore. 97440
Organizing and lobbying: pesticides.
(503) 344-5044

Nuclear Information and Resource
  Service
1346 Connecticut Ave., NW,
4th floor 20036
Information clearinghouse: nuclear
power, alternative energy.
(202) 296-7552

Physicians for Social Responsibility
639 Massachusetts Ave.
Cambridge, Mass. 02139
Research and education: health effects
of nuclear weapons and nuclear war.
(617) 491-2754

Public Media Center
25 Scotland St.,
San Francisco, Calif. 94133
Education: produces media materials,
including print ads, public service announcements, films, light shows, radio
programs and brochures on environmental and other issues.
(415) 434-1403

Rural America
1900 M St., NW, Suite 320 20036
Research and education: health, community development, housing, environmental quality, energy in rural
areas.
(202) 659-2800

Safe Energy Communication Council
1609 Connecticut Ave., NW,
Suite 200 20009
Education, organizing, media access
for anti-nuclear and safe energy activists.
(202) 483-8491

SANE
(National Committee for a SANE
Nuclear Policy)
514 C St., NE 20002
Research, education and lobbying: MX
missile, conversion of defense plants
to peaceful uses, arms reduction, military draft, economic concessions.
(202) 546-7100

Science for the People
897 Main St.,
Cambridge, Mass. 02139
Research and education: radical analysis of current science and technology
issues in agriculture, sociobiology, toxics, pesticides, military.
(617) 547-0370

Sierra Club
530 Bush Street
San Francisco, CA 94108
Lobbying: parks, wilderness, public
lands, air quality, coastal environment,
energy, nuclear waste, toxics.
(415) 981-8634

Southwest Research and
Information Center
P.O. Box 4524, Albuquerque, N.M.
  87106
Research, legal assistance and education: utilities, radioactive waste disposal, energy.
(505) 262-1862

Urban Environment Conference
666 11th St., NW, Suite 1001 20001
Research, education and organizing:
labor, minority, community and environmental issues.
(202) 797-0446

INDEX

Abalone Alliance, 90, 229
Abrams, Floyd, 149
Agent Orange, 17, 35, 67, 135, 226–27
Agribusiness, 38
Agriculture Department (USDA), 74, 166
Allied Chemical Corporation, 19, 77, 149, 166
American Association of University Women, 216
American Friends Service Committee, 99
American Lung Association, 110
Ames test, 48, 50
Amicus curiae (friend of the court), 178
Amoco, 74, 255
Apathy, 151–53
Archer, Victor, 23
Armco Steel Corporation, 71
Asbestos School Hazard Detection and Control Act (1979), 207
Asian-Americans, 209, 210, 211, 212
Association of Community Organizations for Reform Now (ACORN), 110, 117, 135, 172, 180–82, 185, 186, 191, 270, 271
Atomic weapons testing, 159–61
Audubon Society, 99, 119, 146, 225, 252, 256, 260
Axelrod, David, 44, 46

Bailley Alliance, 197
Ballot initiative, 170–71, 172–73
Bechtel Corporation, 67, 71
Bendix Corporation, 228
Bethlehem Steel, 149
Billick, Irwin, 57
Black Hills Ordnance Depot, 36
Blacks, 208–9, 210, 211, 212
Bloomfield, Denny, 204

Boden, Leslie, 79, 80
Booth, Heather, 185
Brathwaite, Helene, 112, 148, 205–7, 220, 248
Brazil, 232
Breit, Luke, 163
Bring Legal Action to Stop Tanks (BLAST), 83–87, 90, 97, 107, 111, 120, 123–24, 126, 144, 145, 196, 220
Bristol Meyers, 231
Brower, David, 255
Brown, Edmund G. (Jerry), 77
Bulloch, McRae, 159
Bunker Hill Company, 23
Burford, Anne Gorsuch, 70, 75, 76, 256, 258
Burton, Phillip, 257
Bush, George, 76

Cadmium poisoning, 239–40
Canvassing, 136
Capitalism, 37, 39, 238, 246, 247
Carey, Hugh, 45
Carson, Rachel, 32, 64, 101
Carter, Jimmy, 45, 76, 78, 145–46, 189, 258
Cashman, Joseph, 73
Champion International Corporation, 164
Chemical Control Corporation, 19–20, 139, 161–62, 214
Chemical Industry Institute of Toxicology, 68
Chemical Manufacturers' Association, 72
Chess, Caron, 134, 173–74
Chevron, 203, 231
Chicanos, 210
Chisso Corporation, 239
Christensen, A. Sherman, 159–60
Circle of Poison, 233

296

Citizen Labor Energy Coalition, 185
Citizens Against Aerial Spraying of Phenoxy Herbicides, 163
Citizens Against Toxic Sprays, 148
Citizen's Alliance, 119, 185
Citizens for a Healthy County, 164–65
Citizens party, 270
Civil disobedience, 186–89
Civil rights movement, 207, 211
Clamshell Alliance, 122–23, 146, 147, 148, 203
Clean Air Act (1963), 64, 69, 76, 196, 256; amendments of (1970), 165–66
Clean Air Coalition, 254
Clean Water Act (1972), 76, 256
Clean Water Act (1977), 166
Cleffi, Vivian, 114, 116
Closing Circle, The (Commoner), 37, 101
Cloward, Richard, 190
Coal, coal mining, 22–23, 25
Coalition for a Reasonable 2, 4-D Policy, 67
Coalition for a United Elizabeth (CUE), 19–21, 119, 129, 138, 139, 161–62, 168, 169, 185, 214
Coalition for the Reproductive Rights of Workers (CRROW), 222, 237
Coalitions, 194–238, 262, 269, 271; formation of, 236–38; international, 230–35; labor movement, 194–205; peace movement, 223–30; people of color in, 205–15; women's movement, 216–23
Coastal Carolina Crossroads, 97–98, 141, 153, 157
Committee for Energy Awareness, 67
Committee of Survivors, 160–61
Committees on occupational safety and health (COSH), 109
Commoner, Barry, 37, 64, 101, 139, 194, 198
Communication networks, international, 234
Community concerns, assessment of, 136–37
Community education, 133–58; goals of, 136; materials for, 138, 143–44; methods of, 138–50, 179; obstacles to, 150–57; planning, 136–38; targeting subgroups in, 137–38
Community forums, 139–40
Community organizations, 135, 248–

52, 269, 271; committee structure of, 129; decentralization of, 130; decision-making in, 120–24; dissent within, 125–26; goal-setting of, 130–31; leadership in, 123–28; national organizations and, 256, 259–60, 261–62; varieties of, 117–19; weaknesses of, 250–52; work allocation in, 128–30; *see also* direct action
Compensation, 267
Compromise, 176, 257–58
Congress, 224, 225; General Accounting Office of, 97, 225; Technology Assessment Office of, 30, 46–47, 85, 97
Consensus process, 121–23
Conservation Foundation, 109, 110
Consolidated Edison, 67, 167
Consumer boycotts, international, 233–34
Consumer organizations, 235
Consumer Product Safety Commission, 231
Consumers' Union, 110
Corporate responsibility, 267–69
Corporate strategies, 251, 258–59
Cosgriff, Edwina, 83–87, 111, 124, 126, 175, 196, 220, 248
Cosgriff, Eugene, 83–87, 248
Cost-benefit analysis, 77–81, 178; cost-effectiveness analysis vs., 266
Council for Countermeasures Against *Itai-Itai* disease, 239–40
Council on Economic Priorities, 175
Council on Environmental Quality, 65
Courtemanche, Verna, 112
Critical Mass, 97
Crocker, George, 192

Daniel, John, 76
Danneker, Gail, 201
Davenport, Elizabeth, 173
DBCP, 34, 139, 231, 268
DDT, 166–67, 209, 231, 261, 268
Defense Department, 35–36, 54, 227
Delaware Valley Toxics Coalition (DVTC), 133–35, 136, 137, 173–74
Denial, 151–55
Deukmejian, George, 163
Diablo Canyon nuclear power plant, 89–90, 186, 255

Dioxin, 15, 20, 29, 39, 67, 73, 77, 102, 162–63, 220, 261
Direct action, 40–41; as demonstration of political power, 184–93; as educational method, 141–43; information needs for, 88–91; limits of, 191–92; regional approach to, 63; tactics for, 191–92
Distrigas Corporation, 83
Doctors for Facts, 164
Doll, Richard, 52–53, 58
Dow Chemical Company, 35–36, 67, 68, 72, 74, 77, 139, 171, 203, 231, 249, 255
DuPont, 55, 57, 67, 71, 228, 230

Earl, Anthony, 226
Ecofeminists, 219, 261
Ecology Action, 192
Educational materials, 138, 143–44
Electoral strategies, 135, 165, 170–79, 257–59, 270–71; guidelines for, 177–79; pros and cons of, 172–77, 190
Electric utilities, 24–25
Electronics Committee on Safety and Health, 199
Eli Lilly, 231
Energy Department, 100–1, 198, 228
Energy Research Foundation (ERF), 229–30
Environmental Action (EA), 173, 252, 254–55, 271
Environmental Defense Fund, 46, 57, 68, 109, 178, 195, 256, 271
Environmental impact report (EIR), 98
Environmental impact statement (EIS), 228
Environmentalists for Full Employment, 201
Environmental movement: program proposed for, 263–69; strategy needed by, 261–63; vision needed by, 251–52, 262
Environmental Pesticide Control Act (1972), 64
Environmental Protection Agency (EPA), 15, 18, 24, 27, 29–31, 33, 34, 42, 44–47, 52, 56, 57, 65, 66, 68–70, 74–77, 80, 81, 84, 93–100, 107, 111, 163, 166, 167, 183, 189, 196, 209, 227, 250, 256–59
Epidemiological studies, 51–53, 104

Epstein, Samuel, 31, 50, 101
Eriksen, Ted, Jr., 163
Export of toxics, 231–33
Extremely-low-frequency (ELF) waves, 111, 224
Exxon Corporation, 19, 53, 57, 71

Fair Share, 103–4, 110, 118–19, 172, 173, 185–86, 189, 190, 270, 271
Fauntroy, Walter, 184
Fay, James, 85
Fazal, Anwar, 233
Fellowship for Reconciliation, 225
Feminists, 216–23; coalition problems and, 221–23; contributions of, 218–21; "radical" vs. "socialist," 219
Fernandes, Jacinta, 20–21, 161–62, 214
Ferrucio, Ken, 183, 184
Florio, James, 74
Folks Organized for Responsible Energy, 140, 173
Food and Drug Act, Delaney Amendment to, 266
Food and Drug Administration, 34, 55, 74
Ford, Gerald R., 78
Ford Motor Company, 149
Foreman, Carol Tucker, 34
Foreman, David, 259–60
Forest Service, 99, 100, 259
Formaldehyde, 30, 39, 68
Formaldehyde Institute, 53, 68
Frayser Health and Safety Committee, 191
Frederick Cancer Research Facility, 57
Freedom of Information Act (1966), 18, 36, 100
Friends of the Earth, 100, 157, 195, 201, 252, 255–56, 257, 260
Froelich, Carol, 114–16, 176, 220
Fuchs, Virgil, 187, 190–91

Garment industry, 200, 209
Geigy Chemical, 166
General Electric Company, 67, 71, 72–73, 186, 199, 228
General Motors, 38, 71
General Tire and Rubber Company, 149
Georgia-Pacific Corporation, 164
Georgia 2000, 171, 173, 177
Germany, Federal Republic of (West

Germany): environmental movement in, 243–47, 249, 252, 261; missiles issue in, 244, 245, 246; nuclear power in, 244
Gibbs, Lois, 43, 45, 104, 117, 125–26, 135, 146, 153, 189, 215, 220, 248
Glenn, John, 204
B. F. Goodrich, 71
Goodyear Atomic, 198
Gorleben live-in, 244
Grantham, Nell, 15–17, 117
Greater Newark Bay Coalition Against Toxic Waste, 20, 98, 103, 110, 119, 138, 139
Green, William, 133
Green Party, West German, 243–47, 257, 258, 259, 270
GTE Sylvania, 225
Guatemala, 231
Guenther, George, 71
Gulf + Western, 149
Gulf South Research Institute, 56

Hagino Noboru, 239–40
Hair, Jay, 254
Haiti, 233
Hanson, Dick, 188
Harvard Medical School, 57
Hazardous Materials Transportation Act (1975), 60–61
Hazardous substances: animal tests for, 49–51, 53; assessment of, 48–53; disposal facilities for, 30–31; illegal dumping of, 71, 92, 98, 99, 102, 147; short-term tests for, 48–49, 51, 53
Hazardous Waste in America (Epstein), 101
Health and Human Services Department (DHHS), 46, 56
Health and safety issues, 194–95
Health surveys, 103–4
Henderson, E. L., 153–54
High-voltage power lines, 26, 187–88
Hill, Bonnie, 102, 111
Hispanics, 208–9, 211, 212
Hooker Chemical Company, 39, 40, 42, 43, 47, 147, 169, 199, 202
Hostility, 151–53, 155–57
Housing and Urban Development Department, 57
Hunt, James, 183–184

Indian Affairs Bureau, 18

Indian Health Service, 17
Industrial Bio-Test Laboratories (IBT), 55–56
Industry: economic policy controlled by, 258–59; legislative influence of, 174–75, 177
Industry-sponsored research, 67–69
Information gathering, 88–113; from government sources, 93–101; from industry, 106–8; from national environmental groups, 109–10; from scientific and medical sources, 101–6
Inner cities, 209
Institute for Food and Development Policy, 233
Interior Department, 258
International Association of Machinists, 196
International Business Machines (IBM), 71, 149
International environmental movement, 230–35
International Ladies Garment Workers Union, 200
International Organization of Consumer Unions, 233
International Paper Corporation, 164
International regulation, 235, 268–69
Intervenor status, 178
Itai-itai disease, 239–40

Japan: compensation for pollution victims in, 243; environmental movement of, 239–43, 247, 249, 252, 258, 261; nuclear power in, 241; peace movement in, 241
Javits, Jacob, 207
Jobs issue, 204–5, 214, 268
Johns-Manville, 76, 149, 169–70
Johnson, Carl, 36
Johnson, Lyndon B., 65, 70
Johnson, Ralph, 161
Justice Department, 55, 56, 71, 76, 100

Kakac, Patty, 188
Kaler, Frank, 267
Kansans for Safe Pest Control, 137
Karch, Nathan, 58
Kelly, Petra, 245
Kennedy, John F., 70
King, Donald, 72

Koch, Edward I., 64
Koons, Charlotte, 216

Laborers' International Union, 196
Labor movement, 194–205; coalition problems in, 196, 200–1; goals of, 199, 201–2; resources of, 108–9, 196–97
Landrigan, Philip, 31
Lavelle, Rita, 77
Lead, lead poisoning, 23, 57, 66, 75, 97, 209
League of Conservation Voters, 65
League of Women Voters, 110, 135, 179
Legal action, 135, 165–70, 177–79; as delaying device, 167–68; drawbacks of, 169–70, 190; guidelines for, 177–79
Lehr, Lewis, 73
Levine, Adeline, 45, 57, 126
Liaison Movement to Prevent Petroleum Development in the Osumi Peninsula, 240–41
Lincoln County Citizens for Common Sense, 164–65
Lincoln County Medical Society, 164
Linton, Rhoda, 218
Liquified natural gas (LNG), 82–87, 90, 97, 120, 144, 196
Lobbyists, 69–70
Long Island Lighting Company, 63, 107, 189
Long Island Safe Energy Coalition, 189
Love Canal, 21, 30, 39, 40, 42–47, 52, 55, 57, 58, 92, 117, 119, 131, 147, 172, 186–89, 213, 220, 256, 258
Love Canal Homeowners' Association (LCHA), 43, 44, 46, 47, 104, 125–26, 130, 136, 145, 147, 150, 189, 202, 213
Lovejoy, Thomas, 261
Lowerey, Joseph, 184
Lucas, Michael, 245
Lyon, Joseph, 160

McCloskey, Michael, 253, 259
McDevitt, George, 200, 202
MacGregor, Gregor, 177
McKean, Margaret, 242
Martinez Environmental Cooperative, 140
MASS BLAST, 86

Mass media, 85, 108; big business links to, 148–50; community education and, 144–50; getting coverage from, 145–47
Mazzocchi, Tony, 194, 203, 205
Meese, Edwin, III, 77
Mexico, 231, 232
Middle Paxton Concerned Citizens, 167–68
Midwest Academy, 127
Miller, Jeff, 76
Miller, Mike, 176
Minamata disease, 239, 240
Mining hazards, 22–24
Minnesota Mining and Manufacturing (3M), 73
Minorities, *see* people of color
Mitsui Mining and Smelting Company, 240
Mobil Oil, 66, 149
Moffett, Toby, 70
Monsanto Company, 56, 57, 66–67, 71, 227, 231
Moran, Jim, 134
Morgan, Karl Z., 160
Mount Sinai Medical Center, 206
Muir, John, 252–53
Multi-issue citizen action groups, 185–86
Multinational corporations, 230–31
Murphy, John, 85–86
Mutsu, 241–42
MX missiles, 227, 228, 254, 258

Nader, Ralph, 110, 119, 133
National Association for the Advancement of Colored People (NAACP), 212–13, 271
National Campaign to Stop the MX, 254
National Cancer Institute, 50, 98
National Citizen's Clearinghouse on Hazardous Waste, 271
National Clean Air Coalition, 109
National Environmental Policy Act (1970), 64, 228, 256
National Institute for Occupational Safety and Health (NIOSH), 23, 26, 31
National liberation movements, 234
National Organization for Women, 216
National organizations, 252–62, 271;

community groups and, 256, 259–60, 261–62; conservation emphasis of, 260–61; electoral activity of, 257–59

National Pollution Discharge Elimination System (NPDES), 93

National Welfare Rights Organization, 195

National Wildlife Federation (NWF), 109, 242, 252, 253–54, 256

National Women's Health Network, 271

Native Americans, 208–9, 210, 211, 212, 219, 248, 261

Natural Resources Defense Council, 178, 195, 256, 260

Navajo Indians, 23, 208

Navy, 223–24, 226

Nestlé Food Corporation, 233

Nevada Nuclear Test Site, 159

New Jersey, 24, 27, 77, 173, 255; Environmental Protection Department of, 19, 20, 103, 115; Health Department of, 115; nuclear freeze ballot referendum of, 229

New Jersey Public Interest Research Group, 92, 93–97

New Mexico Citizens for Clean Air and Water, 167

New Right, 221

New York City, 98; Board of Education of, 203, 207; Department of Environmental Protection of, 32; Fire Department of, 61, 62, 64, 83; Health Department of, 25, 61–63, 97; Police Department of, 64

New York-New Jersey Port Authority, 60, 61–62, 110

New York Public Interest Group, 206

New York State, 24; Environmental Conservation Department of, 30, 42, 98; Health Department of, 43–44, 45–46, 57; Motor Vehicles Department of, 61

Nicaragua, 234

Nixon, Richard M., 71

Nixon, Thelma (Pat), 84

Noble, David, 80

Northwest Coalition Against Pesticide Misuse, 110

Nuclear freeze movement, 228–29, 230

Nuclear industry, 22, 25, 67, 139, 268

Nuclear power, nuclear weapons, 100–1, 142, 150, 216–17, 218, 255–56

Nuclear Regulatory Commission (NRC), 25, 70, 97, 99, 100, 142

Oak Ridge National Laboratory, 25, 153, 160

Occidental Petroleum, 42, 74, 255, 270

Occupational Safety and Health Act (1970), 64

Occupational Safety and Health Administration (OSHA), 65, 70, 71, 74, 75, 78, 196, 227, 231

Oil, Chemical and Atomic Workers' (OCAW) union, 194–95, 196, 197, 198, 203, 205

Olin Corporation, 77, 166

On-site disposal, 267

Ortho, 203

Osumi project, 240–41

Pacific Gas and Electric Company (PGE), 89–90

Paigen, Beverly, 43, 44, 55

Panhandle Environmental Awareness Committee, 228

Peace movement, 204, 223–30; Western European, 245, 247

People of color, 205–15; class oppression of, 210; contributions of, 207–8, 210–12; in high-risk jobs, 208–9; self-determination as goal of, 215

Perpich, Rudy, 73

Pesticides, 22, 32–34, 97, 100, 139, 209, 233

Pesticides Action Network, 233, 234

Peterson, Russell, 256

Peto, Richard, 52–53, 58

Petrochemical industry, 21–22, 28, 37, 203

Petroleum industry, 194–95

Philadelphia Project on Occupational Safety and Health (PhilaPOSH), 133

Philadelphia Public Interest Law Center, 133–34

Physicians for Social Responsibility, 155, 230

Picciano, Dante, 44, 45

Pickett, Elmer, 160

Pilgrim nuclear power plant, 203

302   *Index*

Piven, Frances, 190
Platforms and position papers, 171–72
Political action committees (PACs), 253, 255
Politics of Cancer, The (Epstein), 101
Polychlorinated biphenyls (PCBs), 20, 29, 39, 72, 134, 137, 142, 182–84, 185, 186, 198–99, 208, 211, 261
Polyvinyl chloride (PVC), 29, 39, 51–52, 197
Poor People's Movements (Piven and Cloward), 190
Pope, Carl, 253
Price County (Wisconsin) Nuclear Education Association, 155
Profit motive, 37–40
Project ELF, 223–26, 227
P.S. 208 (New York City), 205–8
Public Interest Research Group, 179
Public Service Company of New Hampshire, 200
Public speaking, 140–41
Puerto Rico: independence movement in, 234; U.S. pharmaceutical companies in, 231

Quinn, John, 82

Racism, 210, 213, 215
Rademacher, John, 70
Rall, David, 45
Rapid City Journal, 36
Rathke, Wade, 191
Reagan, Ronald, 70, 74, 75, 76, 77–78, 100, 109, 196, 203, 227, 244, 252, 253, 254, 258, 271
Reagan administration, 57, 69–70, 75–77, 177, 196, 222, 233, 254, 256, 258
Reed, Scott, 168, 178
Referendums, 170
Regulation, 64–81; government responsibilities for, 265–266; industry influence in, 66–72; resources for, 73–74; state and local, 72–73
Reproductive-rights campaigns, 145, 221–22
Republic Steel Corporation, 71, 75, 255
Reserve Mining Company, 71
Resource Conservation and Recovery Act (1976), 64, 97

Reuben, Melvin, 57
Rights, 261, 264–65
Right-to-know laws, 101, 133–35, 136, 172, 173–74, 197, 201, 203, 248, 269, 271
Right to Life groups, 222
Rockefeller University, 57
Rocky Flats Action Group, 216, 227, 228
Rocky Flats Nuclear Weapons Facility, 35–36, 167, 226–28
Rohm and Haas Company, 71, 133
Rohr Industries, 77
Rollins Environmental Services, 77
Roswell Park Memorial Institute, 43, 44
Rothman, Carol, 228
Ruckleshaus, William, 70, 76, 81
Rusk County Citizen Action Group, 148

Saccharin, 50
Safe Drinking Water Act (1974), 64
Safe Energy Alliance, 144, 203
Safe energy movement, 235
SANE, 225
San Jose State College, 212
Sanjour, William, 31
Savannah River Plant, 229
SCA Corporation, 98, 110
Schecter, Arnold, 29
Schneiderman, Marvin, 58
Schweiker, Richard, 161
Science, scientists, 47–48, 53–59
Scientists' Institute for Public Information, 65
Seabrook nuclear power plant, 122–23, 186, 196, 200
G. D. Searle, 231
Securities and Exchange Commission (SEC), 98, 99, 107
Seneca Army Depot, 36, 204
Sheets, Jim, 196
Shell Oil, 71, 194–95, 197, 198, 203, 231
Shoreham nuclear power plant, 63, 107, 138, 142, 147, 189, 216–17
Shy, Carl, 27
Sierra Club, 99, 109, 117, 119, 146, 178, 195, 201, 212, 225, 242, 252–53, 254, 255, 256, 257, 260
Sierra Leone, 232–33

Silbergeld, Ellen, 46, 57
Silent Spring (Carson), 32, 101
Silkwood, Karen, 217
Silresim Chemical Corporation, 186
Simeral, William, 67
Smith, William D., 224
Smith, William French, 100
Smith Kline, 231
Socialism, 262
Social planning, corporate control of, 249
Solon, Leonard, 25–26
Speak-outs, 139–40, 155, 220
Speicher, Jenny, 225
Stafford, Robert, 258
Standard Oil, 203
State Department, 72
Stauber, John, 225
Stauffer, 180–82, 185
J. P. Stevens, 203
Stockholders' resolutions, 189–90
Stop Project ELF, 111, 138, 144, 223–26
Street theater, 220
Sun Oil, 149
Sweatshops, 200, 209
Synthetic fuel, 25

Taiwan, 232
Tanaka, Kakuei, 240
Teach-ins, 139
Temik, 33–34
Tennesseans Against Chemical Hazards (TEACH), 17, 110, 189
Test ban treaty (1963), 226
Thomas, Lewis, 45
Thomas Panel, 45
Three Mile Island accident, 21, 25, 88, 100, 197, 216, 220
Thunder Hawk, Madonna, 211
Toxic Chemical Task Force, 171
Toxic Substances Control Act (1976), 64, 69, 78, 256
Toyama, Japan, 240
Trampert, Rainer, 246
Transportation, 28, 60–64, 73–74, 137
Transportation Department (DOT), 28, 60–64, 73–74
Tripp, John, 191
Twin Cities Army Ammunition Plant, 36
2,4-D, 17, 67, 68, 99, 162

2,4,5-T, 68, 99, 102, 106, 162, 249
Two-party system, 259, 270

Udall, Stewart, 70
Union Carbide, 33, 55, 66, 153, 228
Union of Concerned Scientists, 64–65
United Church of Christ Commission for Racial Justice, 271
United Mine Workers, 203
United Steel Workers of America, 196
United Technologies, 66
Upjohn, 55
Uranium, uranium mining, 17–18, 23–24, 105, 140, 211
Urban Environment Conference, 110, 271
Utility rates organizations, 235

Velsicol Chemical Company, 15–17, 55, 70, 168, 189, 231, 232, 270
Vogel, Robert, 133

Wallick, Frank, 196
Ward, Yolanda, 217
Ward Transformer Company, 183
Warner Lambert, 231
Warnock, Donna, 217, 218
Warren, Jacqueline, 68
Warren County Citizens Concerned About PCBs, 183
Washington Heights Health Action Project, 60, 62–63, 97, 174, 200, 214
Watt, James, 253, 256, 258
Weiner, Ruth, 81
Westinghouse Electric, 149
West Virginia Citizens Against Toxic Sprays, 143
Westway project, 111–12, 204
We Who Care, 114–16, 123, 125, 141, 147, 155, 176
Weyerhauser, 70, 74, 76, 164, 255
Whalen, Robert, 42–43
White, Leon, 183
Wilderness Society, 195, 252, 259
Williams, Margaret, 213
Wilson, David, 146, 169
Winpisinger, William, 197
Women of All Red Nations (WARN), 17–19, 119, 211, 248
Women Opposed to Nuclear Technology (WONT), 216–17, 219, 223
Women's Health Alliance, 216

Women's International League for Peace and Freedom, 216

Women's movement, 216–23

Women's Pentagon Action, 216, 217–18, 219, 223

World Health Organization, 234

World Wildlife Fund, 261

Wright, Paul, 56

Yannacone, Victor, 167

Zero exposure, 266

Zimmerman, Burke, 80–81